住房和城乡建设部"十四五"规划教材
国际化人才培养土木工程专业系列教材

Foundation Engineering
基 础 工 程

Tong Zhaoxia, Feng Jinyan, Du Bowen
童朝霞　冯锦艳　杜博文　主编

中国建筑工业出版社

图书在版编目（CIP）数据

基础工程 = Foundation Engineering：英文 / 童朝霞，冯锦艳，杜博文主编. — 北京：中国建筑工业出版社，2023.12

住房和城乡建设部"十四五"规划教材 国际化人才培养土木工程专业系列教材

ISBN 978-7-112-29764-1

Ⅰ.①基… Ⅱ.①童… ②冯… ③杜… Ⅲ.①基础（工程）－高等学校－教材－英文 Ⅳ.①TU47

中国国家版本馆 CIP 数据核字(2024)第 078162 号

责任编辑：聂 伟 王 跃
责任校对：张惠雯

住房和城乡建设部"十四五"规划教材
国际化人才培养土木工程专业系列教材
Foundation Engineering
基础工程
Tong Zhaoxia，Feng Jinyan，Du Bowen
童朝霞 冯锦艳 杜博文 主编
*
中国建筑工业出版社出版、发行（北京海淀三里河路 9 号）
各地新华书店、建筑书店经销
北京红光制版公司制版
天津画中画印刷有限公司
*
开本：787 毫米×1092 毫米 1/16 印张：18¼ 字数：456 千字
2024 年 1 月第一版 2024 年 1 月第一次印刷
定价：58.00 元（赠教师课件）
ISBN 978-7-112-29764-1
（42182）

版权所有 翻印必究
如有内容及印装质量问题，请联系本社读者服务中心退换
电话：(010) 58337283 QQ：2885381756
（地址：北京海淀三里河路 9 号中国建筑工业出版社 604 室 邮政编码：100037）

This book is the 14th Five Year Planned Textbook of the Ministry of Housing and Urban-Rural Development of the People's Republic of China. Based on the latest codes on the survey, design, and construction of foundation engineering, it systematically introduces the design principles and methods of foundation engineering.

This book includes ten chapters, which are introduction, site exploration and characterization, design of shallow foundation on natural subgrade, continuous foundations, pile foundations, retaining and protection structures for foundation excavation, ground improvement, special soil subgrade, seismic analysis and design of foundations, and comparison for foundation engineering design codes.

This book can be used as a teaching material for civil engineering majors in higher education institutions, as well as technical reference for engineers engaged in foundation engineering.

本书为住房和城乡建设部"十四五"规划教材，根据我国基础工程相关的勘察、设计和施工等最新规范编写，系统介绍了基础工程的设计原理和方法。

全书共分10章，包括引言、地基勘察、天然地基上的浅基础设计、连续基础、桩基础、基坑支护结构、地基处理、特殊土地基、地基抗震分析和设计，以及与其他国家基础工程设计规范对比。

本书可作为高等学校土木工程专业的教学用书，也可供其他专业师生以及从事基础工程设计和施工的技术人员参考。

为了更好地支持相应课程的教学，我们向采用本书作为教材的教师提供课件，有需要者可与出版社联系。建工书院：http://edu.cabplink.com，邮箱：jckj@cabp.com.cn，2917266507@qq.com，电话：(010) 58337285。

出 版 说 明

党和国家高度重视教材建设。2016年，中办国办印发了《关于加强和改进新形势下大中小学教材建设的意见》，提出要健全国家教材制度。2019年12月，教育部牵头制定了《普通高等学校教材管理办法》和《职业院校教材管理办法》，旨在全面加强党的领导，切实提高教材建设的科学化水平，打造精品教材。住房和城乡建设部历来重视土建类学科专业教材建设，从"九五"开始组织部级规划教材立项工作，经过近30年的不断建设，规划教材提升了住房和城乡建设行业教材质量和认可度，出版了一系列精品教材，有效促进了行业部门引导专业教育，推动了行业高质量发展。

为进一步加强高等教育、职业教育住房和城乡建设领域学科专业教材建设工作，提高住房和城乡建设行业人才培养质量，2020年12月，住房和城乡建设部办公厅印发《关于申报高等教育职业教育住房和城乡建设领域学科专业"十四五"规划教材的通知》（建办人函〔2020〕656号），开展了住房和城乡建设部"十四五"规划教材选题的申报工作。经过专家评审和部人事司审核，512项选题列入住房和城乡建设领域学科专业"十四五"规划教材（简称规划教材）。2021年9月，住房和城乡建设部印发了《高等教育职业教育住房和城乡建设领域学科专业"十四五"规划教材选题的通知》（建人函〔2021〕36号）。为做好"十四五"规划教材的编写、审核、出版等工作，《通知》要求：（1）规划教材的编著者应依据《住房和城乡建设领域学科专业"十四五"规划教材申请书》（简称《申请书》）中的立项目标、申报依据、工作安排及进度，按时编写出高质量的教材；（2）规划教材编著者所在单位应履行《申请书》中的学校保证计划实施的主要条件，支持编著者按计划完成书稿编写工作；（3）高等学校土建类专业课程教材与教学资源专家委员会、全国住房和城乡建设职业教育教学指导委员会、住房和城乡建设部中等职业教育专业指导委员会应做好规划教材的指导、协调和审稿等工作，保证编写质量；（4）规划教材出版单位应积极配合，做好编辑、出版、发行等工作；（5）规划教材封面和书脊应标注"住房和城乡建设部'十四五'规划教材"字样和统一标识；（6）规划教材应在"十四五"期间完成出版，逾期不能完成的，不再作为《住房和城乡建设领域学科专业"十四五"规划教材》。

住房和城乡建设领域学科专业"十四五"规划教材的特点，一是重点以修订教育部、住房和城乡建设部"十二五""十三五"规划教材为主；二是严格按照专业标准规范要求编写，体现新发展理念；三是系列教材具有明显特点，满足不同层次和类型的学校专业教学要求；四是配备了数字资源，适应现代化教学的要求。规划教材的出版凝聚了作者、主审及编辑的心血，得到了有关院校、出版单位的大力支持，教材建设管理过程有严格保障。希望广大院校及各专业师生在选用、使用过程中，对规划教材的编写、出版质量进行反馈，以促进规划教材建设质量不断提高。

住房和城乡建设部"十四五"规划教材办公室
2021 年 11 月

Preface

Foundation engineering is a discipline that studies the subgrade and foundation issues during the design and construction of buildings. It is one of the main courses for civil engineering majors. Foundation engineering is related to a wide range of disciplines, such as soil mechanics, engineering geology, material mechanics, structural mechanics, reinforced concrete structures, engineering construction and other disciplines. It involves extensive contents. With the development of science and technology, the theory and technology of foundation engineering are constantly changing.

This textbook is the 14th Five Year Planned Textbook of the Ministry of Housing and Urban-Rural Development of the People's Republic of China. Based on the latest codes on the survey, design, and construction of foundation engineering, it systematically introduces the design principles and methods of foundation engineering. Meanwhile, the comparison for foundation engineering design codes is included aiming to cultivate high-level technical talents with international competitiveness.

This textbook includes ten chapters, which are introduction, site exploration and characterization, design of shallow foundation on natural subgrade, continuous foundations, pile foundations, retaining and protection structures for foundation excavation, ground improvement, special soil subgrade, seismic analysis and design of foundations, and comparison for foundation engineering design codes. Among them, Chapters 1, 2, 6, 9 and 10 were written by Tong Zhaoxia. Chapters 4, 7 and 8 were written by Feng Jinyan. Chapters 3 and 5 were written by Du Bowen. Ochieng J. Obongo proofread Chapters 2, 3 and 5. The entire book was compiled by Tong Zhaoxia.

Limited to the authors' level, there are many deficiencies and errors in this textbook. Readers are welcome to criticize and correct it.

Editors

July, 2023

前　言

　　基础工程是研究建筑物在设计和施工中与地基和基础问题有关的学科，是土木工程专业的主干课程。基础工程涉及的学科很广，包括土力学、工程地质、材料力学、结构力学、钢筋混凝土结构、工程施工等学科领域，内容广泛，综合性强。随着科学技术的发展，基础工程的理论和技术日新月异。

　　本书为住房和城乡建设部"十四五"规划教材，依据我国基础工程相关的勘察、设计和施工等最新规范编写，系统介绍了基础工程的设计原理和方法。并与其他国家基础工程设计规范进行了对比，以培养具有国际竞争力的高水平技术人才。

　　全书共分10章，包括引言、地基勘察、天然地基上的浅基础设计、连续基础、桩基础、基坑支护结构、地基处理、特殊土地基、地基抗震分析和设计，以及与其他国家基础工程设计规范对比。其中第1、2、6、9、10章由童朝霞编写，第4、7、8章由冯锦艳编写，第3、5章由杜博文编写，Ochieng J. Obongo对第2、3、5章进行了校对，全书由童朝霞统稿。

　　限于作者水平，书中定有欠妥和错误之处，敬请读者批评指正。

编者

2023年7月

Contents

Chapter 1　Introduction ·· 1
  1.1　Contents of foundation engineering ··· 1
  1.2　Importance of foundation engineering ·· 2
  1.3　Development of foundation engineering ··· 4
  1.4　Curriculum features and requirements ·· 6
    Questions ·· 6

Chapter 2　Site Exploration and Characterization ································· 7
  2.1　Introduction ·· 7
  2.2　Geological exploration ·· 8
    2.2.1　Geological geophysical prospecting ··· 8
    2.2.2　Subsurface exploration and sampling ··· 9
    2.2.3　Exploratory trenches ··· 13
  2.3　In-situ testing ·· 13
    2.3.1　Plate load test (PLT) ··· 14
    2.3.2　Cone penetration test (CPT) ··· 16
    2.3.3　Standard penetration test (SPT) ·· 18
    2.3.4　Dynamic cone penetration test (DCPT) ······································· 20
    2.3.5　Vane shear test (VST) ··· 21
  2.4　In-situ monitoring ·· 22
    2.4.1　Structure settlement measurement ·· 23
    2.4.2　Excavation pit monitoring ·· 24
    2.4.3　Groundwater monitoring ·· 25
  2.5　Main contents of the geotechnical investigation report ······················ 25
    Questions ·· 26

Chapter 3　Design of Shallow Foundation on Natural Subgrade ··········· 28
  3.1　Introduction ·· 28
    3.1.1　Design grade of foundation ·· 28
    3.1.2　Design principle of foundation ··· 29
    3.1.3　Design loads ··· 31
    3.1.4　Steps in the design of the shallow foundation ····························· 31
  3.2　Types of the shallow foundation ··· 32
  3.3　Depth of foundations ·· 36

  3.3.1 Function, type and loads of buildings ································ 37
  3.3.2 Engineering geological and hydrogeological conditions ················ 38
  3.3.3 Influence of adjacent buildings ···································· 39
  3.3.4 Frost heave of subgrade ·· 40
 3.4 Subgrade bearing capacity verification ··································· 45
  3.4.1 Subgrade bearing capacity ·· 45
  3.4.2 Bearing capacity verification of bearing soil layer ···················· 47
  3.4.3 Bearing capacity verification of weak substratum ···················· 49
 3.5 Evaluation of subgrade settlement ······································ 52
  3.5.1 Allowable subgrade deformation ···································· 52
  3.5.2 Calculation of subgrade settlement ································ 55
 3.6 Evaluation of subgrade stability ·· 57
 3.7 Design of unreinforced spread footings ·································· 60
 3.8 Design of spread footings ··· 61
  3.8.1 Pad foundations ·· 61
  3.8.2 Strip footings under wall ·· 65
  3.8.3 Structural requirement of spread footings ·························· 65
 3.9 Methods to reduce damages due to uneven settlement of buildings ············ 67
  3.9.1 Design methods ·· 68
  3.9.2 Structural methods ·· 70
  3.9.3 Construction methods ·· 72
  Questions ·· 73

Chapter 4 Continuous Foundations ································ 75
 4.1 Introduction ·· 75
 4.2 Interaction of subgrade, foundation and superstructure ···················· 76
  4.2.1 Interaction concept of subgrade, foundation and superstructure ········ 76
  4.2.2 Subgrade models ·· 80
 4.3 Strip footings under columns ·· 82
  4.3.1 Structure requirement and construction ···························· 82
  4.3.2 Internal force calculation ·· 85
  4.3.3 Cross foundations under columns ·································· 88
 4.4 Raft foundations and box foundations ··································· 89
  Questions ·· 92

Chapter 5 Pile Foundations ·· 94
 5.1 Introduction ·· 94
 5.2 Classification and selection of piles ····································· 96
  5.2.1 Classification of piles ·· 96
  5.2.2 Pile type selection ·· 100

5.3　Load transfer of vertical bearing piles ······ 101
 5.3.1　Load transfer mechanism ······ 101
 5.3.2　Side resistance of piles ······ 103
 5.3.3　Pile tip resistance ······ 104
5.4　Vertical bearing capacity of piles ······ 106
 5.4.1　Vertical bearing capacity of a single pile ······ 106
 5.4.2　Pile group effect ······ 109
 5.4.3　Vertical bearing capacity of a composite foundation pile ······ 111
5.5　Vertical bearing capacity verification of pile foundations ······ 113
 5.5.1　Load applied on a foundation pile ······ 113
 5.5.2　Vertical bearing capacity verification of pile foundations ······ 114
5.6　Negative friction resistance of piles ······ 116
 5.6.1　Concept of negative friction of piles ······ 116
 5.6.2　Distribution of negative friction ······ 116
 5.6.3　Calculation of negative friction ······ 118
 5.6.4　Bearing capacity verification of negative friction piles ······ 119
5.7　Uplift bearing capacity verification of piles ······ 120
 5.7.1　Uplift bearing capacity of piles ······ 120
 5.7.2　Verification of uplift bearing capacity of piles ······ 121
5.8　Lateral bearing capacity verification of piles ······ 122
 5.8.1　Behavior of piles under lateral loads ······ 122
 5.8.2　Static lateral load tests of a single pile ······ 123
 5.8.3　Theoretical analysis of elastic long pile under lateral loads ······ 125
 5.8.4　Lateral bearing capacity verification ······ 126
5.9　Settlement of pile foundations ······ 127
 5.9.1　Equivalent deep foundation method ······ 127
 5.9.2　Mindlin-Geddes Method ······ 129
5.10　Structure design of pile caps ······ 131
 5.10.1　Basic requirements for the structure of pile caps ······ 131
 5.10.2　Bending calculation of pile caps ······ 133
 5.10.3　Punching calculation of independent pile caps under columns ······ 135
 5.10.4　Shear calculation of independent pile cap of pile foundation ······ 139
 Questions ······ 140

Chapter 6　Retaining and Protection Structures for Foundation Excavation ······ 142

6.1　Introduction ······ 142
6.2　Foundation excavation and supporting methods ······ 144
 6.2.1　Safety grade of foundation excavation and retaining structures ······ 144

6.2.2	Types of foundation excavation and retaining structures	145
6.3	Calculation of earth and water pressure on retaining structures	152
6.4	Stability evaluation of foundation excavation	153
6.4.1	Stability of pile or wall retaining structure	153
6.4.2	Stability of gravity cement-soil wall	159
6.4.3	Stability of soil nailing wall	160
6.5	Design of pile and wall retaining structure	164
6.5.1	Internal force and deformation of pile or wall retaining structure	164
6.5.2	Anchor and strut support	166
	Questions	170

Chapter 7 Ground Improvement … 171

7.1	Introduction	171
7.1.1	Type and characteristics of soft soil	171
7.1.2	Purpose and requirements of ground improvement	173
7.1.3	Classification of ground improvement methods	173
7.2	Replacement cushion method	176
7.2.1	Function of replacement cushion	177
7.2.2	Design of replacement cushion	178
7.3	Preloading method	181
7.3.1	Mechanism of various preloading methods	181
7.3.2	Design of surcharge preloading method	183
7.3.3	Design of vacuum preloading method	187
7.4	Dynamic compaction	189
7.4.1	Strengthening mechanism of dynamic compaction	189
7.4.2	Design of dynamic compaction	190
7.5	Composite foundation	193
7.5.1	Concept and classification of composite foundation	193
7.5.2	Reinforcement mechanism of composite foundation	195
7.5.3	Design parameter of composite foundation	196
7.5.4	Bearing capacity of composite foundation	197
7.5.5	Settlement calculation of composite foundation	200
7.5.6	Sand-gravel pile composite foundation	201
7.5.7	Cement-soil mixed pile composite foundation	203
7.6	Grouting reinforcement ground	207
7.6.1	Introduction of grouting method	207
7.6.2	Classification of grouting method	207
7.7	Geosynthetics reinforcement ground	209
7.7.1	Types of geosynthetics	209

7.7.2 Functions of geosynthetics	212
Questions	214

Chapter 8 Special Soil Subgrade — 215

8.1 Collapsible loess subgrade — 215
   8.1.1 Main characteristics and distribution of loess — 215
   8.1.2 Collapsibility of loess and its influencing factors — 218
   8.1.3 Collapsibility evaluation of loess — 220
   8.1.4 Engineering measures for collapsible loess subgrade — 225

8.2 Expansive soil subgrade — 227
   8.2.1 Characteristics of expansive soil and its harm to buildings — 227
   8.2.2 Characteristic parameters of expansive soil — 229
   8.2.3 Key points for the foundation design of expansive soil site — 232
   8.2.4 Engineering measures for expansive soil subgrade — 238
   Questions — 240

Chapter 9 Seismic Analysis and Design of Foundations — 242

9.1 Introduction — 242

9.2 Site and subgrade — 245
   9.2.1 Site classification — 245
   9.2.2 Subgrade liquefaction assessment — 248

9.3 Seismic bearing capacity verification of shallow foundations — 251
   9.3.1 Seismic resistance standards — 251
   9.3.2 Seismic bearing capacity of natural subgrade — 252

9.4 Seismic bearing capacity verification of pile foundations — 254
   9.4.1 Damages of pile foundation under earthquake — 254
   9.4.2 Seismic bearing capacity of pile foundation — 256

9.5 Seismic measures for foundations — 258
   Questions — 261

Chapter 10 Comparison for Foundation Engineering Design Codes — 262

10.1 Introduction to design codes in different countries — 262
   10.1.1 Introduction to Eurocodes — 263
   10.1.2 Introduction to USA Standards — 265

10.2 Comparison of the bearing capacity verification for shallow foundations — 265
   10.2.1 Comparison of bearing pressure calculation — 265
   10.2.2 Comparison of bearing capacity calculation — 267

10.3 Comparison of vertical bearing capacity calculation of piles — 270
   Questions — 277

References — 278

Chapter 1   Introduction

1.1   Contents of foundation engineering

All the buildings and structures, such as houses, roads, bridges, dams, oil tanks, etc., are located on the ground. As Figure 1-1 shown, they generally consist of three parts: the superstructure, foundation and subgrade. The bottom part of structures is called the foundation which transfers the load from the upper structure to the soil. The part of the soil affected by the upper structure load is called the subgrade on which the additional stress and deformation will be produced.

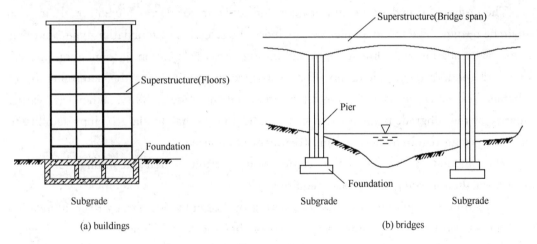

Figure 1-1   Composition of buildings and bridges

When the subgrade is composed of several layers of soil, the part of the soil layer that directly contacts the bottom surface of the foundation and bears the main load is called the bearing layer, and the other soil layers below the bearing layer are called the underlying layer. When the soil of the bearing layer and the underlying layer has good mechanical properties, the requirements of the upper structure for the strength, deformation, and stability of the foundation are easily met. The subgrade can be divided into two types: natural subgrade and artificial subgrade. The natural subgrade is composed of natural soil layers which can meet the design requirements without treatment, such as dense sand layers. The artificial subgrade is composed of soil layers which has undergone manual treatment to meet the requirements, such as soft clay. It is clear that using natural subgrade is good and the most economical when the design requirements can be met for natural subgrade. Meanwhile, according to the embedding depths of the foundation, the foundation can also be divided into two types: shallow foundation and deep foundation. Shallow foundations are usually embedded at a depth of less than 5 m and can be constructed using simple construction processes. The spread foundations and strip foundations under columns belong to shallow foundation. However, if the foundation is deeply buried and can only be constructed using special construction methods, it is called the deep foundation, such as pile foundation, pier foundation, open caisson.

Foundation engineering includes the design, construction and monitoring of foundations and subgrade. Some parts such as reinforcement calculation for pad foundations under columns and construction techniques for shallow foundations have already been covered in courses such as concrete structure and civil engineering construction. The parts which are closely related to geotechnical engineering, such as foundation embedment depth, foundation bearing capacity, foundation deformation calculation, stability of foundation pit and ground treatment, will be discussed in detail in this course.

The design of foundation engineering includes two major parts: foundation design and subgrade design. The foundation design includes the selection of foundation type, determination of foundation embedment depth and bottom area, calculation of foundation internal force and reinforcement calculation. The subgrade design includes the determination of subgrade bearing capacity, subgrade deformation and subgrade stability. When the bearing capacity of the subgrade is insufficient or the subgrade soil has too large compressibility to meet the design requirements, ground treatment is required.

The function of the foundation engineering determines that the foundation design must meet the following three basic requirements:

(1) Strength requirements: the load acting on the subgrade through the foundation shall not exceed its bearing capacity. It ensures that the foundation will not be damaged due to the shear stress in the subgrade soil exceeding its strength.

(2) Deformation requirements: the deformation of the foundation does not exceed the allowable deformation of the building. It ensures that the upper structure is not damaged or affected by excessive deformation.

(3) Other requirements for the upper structure: in addition to meeting the above two requirements, the foundation should also meet the strength, stiffness and durability requirements of the upper structure.

1.2 Importance of foundation engineering

Foundation engineering is important in civil engineering. To solve engineering problems related to foundation and subgrade, foundation engineering requires to use theories and methods of geotechnical engineering. Meanwhile, due to its unique functions and structural requirements, foundation engineering requires a large number of structural calculations in foundation design, so foundation engineering is also closely related to structural design and calculation. Foundation engineering is closely related to geotechnical engineering, structural engineering and construction engineering.

The design and construction of the foundation are an essential part of the whole structure design and construction. When the subgrade soil conditions are complex or soft, the foundation design often becomes the most difficult and the first problem to be solved in engineering construction. Meanwhile, due to the complexity and variability of geotechnical conditions, as well as the limitations of survey work, the uncertainty of geotechnical engineering often makes foundation engineering problems the most challenging.

Foundation engineering is the basis of a building and directly relates to the stability of the upper structure. As far as the engineering cost is concerned, the ratio of foundation cost to the total cost of the whole projects is increasing with the utilization of complex geology, and it reaches as high as 30%. The correct survey, design and construction of the foundation can not only ensure the successful completion of the whole project, but also save the project investment. On the contrary, there are also many foundation instability accidents caused by excessive or uneven deformation of the foundation, insufficient strength of the subgrade, or problems in foundation design and construction. A famous failure case is the Transcona Grain Elevator in Canada, which was shown in Figure 1-2. The plan of the Transcona Grain Elevator is rectangular, with a length of 59.44m, a width of 23.47 m and a height of 31.0 m. Its foundation adopted a reinforced concrete raft foundation, with a thickness of 61 cm and a buried depth of 3.66 m. The construction of the Transcona Grain Elevator began in 1911 and was completed in the autumn of 1913. Starting from September 1913, grain was loaded and carefully distributed. In October, when 31822 m^3 of grain was loaded, it was found that the vertical settlement reached 30.5 cm within one hour. The Transcona Grain Elevator tilted westward and collapsed within 24 hours, with a tilt of 26.53° away from the perpendicular. The western end of the Transcona Grain Elevator sank 7.32 m, and the eastern end lifted 1.52 m. After repair, it is still 4 m lower than before.

Foundation engineering is very important in civil engineering. Due to its underground invisible characteristics, most accidents in engineering construction are caused by foundation problems. Therefore, the survey, design and construction quality of foundation engineering are directly related to the safety of the upper structure. The foundation engineering should be carefully designed and constructed.

Figure 1-2 Transcona Grain Elevator in Canada

1.3 Development of foundation engineering

Foundation engineering is not only an old engineering technology, but also a young applied science. Generally, the practical application of foundation engineering is ahead of theoretical research.

In ancient times, our ancestors have created exquisite techniques in the previous construction activities. For example, the Hemudu site found on the south bank of the Qiantang River in China has wooden piles, which were driven into the swamp 7000 years ago. In the Song Dynasty, Caixiang built Quanzhou Wan'an stone bridge on the Luoyang River where the water was deep and the river was flowing fast. The foundation of the bridge was fixed with oysters, forming a similar raft foundation with a width of 25 m and a length of 1000 m. In the early Northern Song Dynasty, the carpenter Yu Hao built the wooden tower of Kaibao Temple in Kaifeng (989 AD). Due to the prevailing northwest wind, the body of Kaibao Temple built on saturated soil was tilted to the northwest to overcome uneven settlement of the foundation with the help of long-term wind force. In addition, China's world-renowned projects, such as the Great Wall, the Sui Dynasty Grand Canal and the Zhaozhou Stone Arch Bridge, have all survived numerous strong earthquakes and strong winds due to their solid foundation. The traditional foundation construction methods, such as building foundation with stones, making pile foundation with wood, lime soil cushion or shallow foundation, filling and compaction, are still used in some projects.

From the 18th century to the 19th century, people encountered many problems related to soil mechanics in large-scale construction, which promoted the development of soil mechanics. In 1973, French scientist C. A. Coulomb proposed the formula for shear strength of sand and the sliding wedge theory for soil pressure on retaining walls. In 1857, British scientist W. J. M. Rankine proposed the classical earth pressure theory from another perspective. In 1856, French engineer H. Darcy proposed the Darcy's law of laminar flow. In 1885, French scientist J. Boussinesq proposed a theoretical solution for the stress and displacement of a semi-infinite elastic body under vertical concentrated loads. The work of these pioneers laid a solid foundation for the establishment of soil mechanics. In 1925, Terzaghi, an American scientist, published the first publication in soil mechanics in which the basic theories and methods of soil mechanics and foundation engineering are systematically discussed. It promoted the rapid development of soil mechanics. In 1948, Terzaghi and R. Peck published *Engineering Practical Soil Mechanics* which combined theoretical testing with engineering experience and promoted the development of soil mechanics and foundation engineering.

In 1936, the first International Soil Mechanics & Foundation Engineering Conference was held in Harvard, the USA. Soil Mechanics and Foundation Engineering developed rapidly as an independent science. After the 1970s, the international conference revised Soil Mechanics & Foundation Engineering into Geotechnique. Soil mechanics is the theoretical basis of the discipline, which studies the mechanical characteristics, stress-strain relationship, strength and seepage of geotechnical materials. The foundation engineering is to solve the technical problems of structures on geotechnical materials. Soil mechanics and foundation engineering are the integration of theory and engineering application.

In the application technology of foundation engineering, due to the construction of super high-rise buildings, large steel plants, offshore oil platform, large concrete storage tank, artificial island, etc., the foundation technology has obtained significant development. For example, the bearing capacity of a single pile can reach tens of thousands of tons, and the maximum diameter of a cast-in-place pile can reach several meters, with a depth exceeding 100 m. The buried depth of pile foundation of Jin Mao Tower in Shanghai has reached more than 80 m. The pile length of Suzhou Bridge has reached about 120 m, and the diameter of a single pile of Shaojia Passage has reached 3.8 m. Many new piles such as steel pipe piles large steel piles, prestressed concrete pipe piles, DX expansion piles and reinforced cement-soil mixing piles are being widely used. Meanwhile, the design theory of pile foundation has also been greatly developed and applied, such as the composite pile foundation theory considering the load shared by pile and soil.

With the rapid development of China, the requirements for the use of foundation engineering have increased and the application is becoming more and more diversified. The discipline of soil mechanics and foundation engineering in China will be updated and develop faster. It will gradually transfer to intelligent geotechnical engineering.

1.4 Curriculum features and requirements

This textbook consists of ten chapters which mainly introduce shallow foundation, continuous foundation, pile foundation, retaining and protection structures for foundation excavation, ground improvement, special soil subgrade seismic resistance of foundations, and comparison of foundation engineering codes. It belongs to an engineering discipline which introduces the design and construction techniques of foundations on rock and soil layers. This textbook is based on soil mechanics and involves engineering geology, construction materials, reinforced concrete, seismic resistance and other disciplines. The following key points should be highlighted.

(1) Basic theories and concepts of foundation engineering

Students should understand types and characteristics of various foundations, as well as the basic principles of foundation design and calculation. They should be able to combine relevant mechanics and structural theory to analyze and solve foundation problems.

(2) Site exploration

Students should understand the basic principles and operating techniques of main site exploration tests. They should grasp the methods to determine the bearing capacity of various foundations and use the indoor and on-site tests to solve some geotechnical problems.

(3) Local practice

Due to the complexity of the foundation engineering, it is difficult to find identical examples in foundation engineering. Therefore, the theory should be used by combining with local practice.

(4) Use of relevant codes

The design and construction of foundation engineering mustbe conducted according to the related codes. In China, different industries have different codes, such as *Code for Design of Building Foundation* (GB 50007—2011), *Specifications for Design of Foundation of Highway Bridge and Culverts* (JTG 3363—2019), and *Technical Code for Building Pile Foundations* (JGJ 94—2008). Due to the different application objects, there are significant differences in certain aspects among these codes. Various codes should not be mixed or misused.

Questions

1-1　What is a foundation and what is the subgrade? Please give the differences and connections between foundations and subgrade.

1-2　What is the shallow foundation? What is the deep foundation?

1-3　What is the natural subgrade and what is the artificial subgrade?

Chapter 2 Site Exploration and Characterization

2.1 Introduction

Engineers need to know the properties of the materials they work with. In geotechnical engineering, engineers work with rocks and soil which are materials with unknown properties. Therefore, it is necessary for geotechnical engineers to obtain the information about subsurface conditions at the site proposed for construction.

Site exploration involves site investigation and/or soil exploration which involves obtaining information about the subsurface conditions at the proposed construction site. It is important for the geotechnical engineer to be familiar with the technical know-how of obtaining this information and know how to use it to achieve the required precision and accuracy associated with it.

Definitions associated with site exploration include the geological exploration and the in-situ testing. Geological exploration involves assessing the geological qualities of an area to determine what may be present in the rock. In-situ testing involves testing methods conducted in or on site.

The followings are the objectives of site exploration.

(1) Determining the order of occurrence of soil and rock strata.

(2) Determining the location of ground water table level and its applications.

(3) Determining the engineering properties of soil.

(4) Determining the probable and maximum differential settlements.

(5) Finding the bearing capacity of soil.

(6) Predicting the lateral earth pressure against retaining walls.

(7) Determining suitable soil improvement techniques.

(8) Selecting suitable construction equipment.

(9) Forecasting problems occurring in foundations and their solutions.

Depending on the size of the project, site exploration may require a two-phased or three-phased exploration procedure.

Phase 1—Literature review and site reconnaissance

This phase includes the following steps:

—reviewing geological reports and maps, topographic maps, aerial photographs, previous subsurface investigation reports and soil survey reports, etc. ;

Chapter 2 Site Exploration and Characterization

—visiting the site to observe access, current land use and surface features such as topography, outcrops, stability and drainage features,etc. ;

—collecting project information such as performance requirements and special features,etc..

Phase 2—Preliminary field investigations and laboratory testing

The aim of this phase is to develop the general stratigraphy and obtain the soil and rock characteristics of the site. The field investigation provides information of what type of soil and rocks may be present on the surface and surrounding area that may affect construction or excavation that may be required later during construction. It also aids the engineer to choose which sampling methods are most suitable for the site.

After sampling, laboratory tests are conducted to understand the physical properties of the soil such as shear and compressive strength. These factors directly affect the type of foundation that is required for construction.

2.2 Geological exploration

Geological exploration in foundation engineering refers to the process of investigating the physical and geological characteristics of a site before constructing a foundation. The purpose of geological exploration is to determine the soil and rock properties of the site, which helps in designing the foundation and selecting the appropriate type of foundation.

The geological exploration process typically involves field investigations, laboratory tests, and analysis of the data collected. The field investigations may include drilling, soil sampling, and geophysical surveys to obtain information about the soil and rock structure, the depth of the soil layers, the groundwater level, and other important factors. The laboratory tests are conducted on the soil and rock samples to determine their physical and mechanical properties.

The data collected from the geological exploration is used to design the foundation, considering factors such as the load-bearing capacity of the soil and rock, the settlement characteristics of the soil, and the potential for soil liquefaction or landslides. This information is critical to ensure that the foundation is stable, safe and durable.

2.2.1 Geological geophysical prospecting

In foundation engineering, the main objective of geological geophysical prospecting is to investigate the subsurface geology of a site to determine the properties of the soil and rock, and to help design the foundation accordingly.

One of the advantages of geophysical prospecting is that it can provide information on the subsurface geology without the need for extensive drilling and excavation. This can save time and money in exploration and environmental studies.

Seismic reflection is one of the most commonly used geophysical methods as shown in Figure 2-1. It involves the use of sound waves to create images of the subsurface structures. The sound waves are generated by an energy source such as an airgun, and the waves are reflected back to the surface by the subsurface rocks and sediment layers. The reflected waves are recorded by receivers which are called geophones, and the data is used to create images of the subsurface structures.

Electrical resistivity is another geophysical method commonly used in groundwater and geological exploration. It involves passing an electric current through the ground and measuring the resistance of the soil or rock to the flow of electricity. Different soil and rock types have different electrical resistivities, which can provide information on the subsurface geology and the presence of groundwater.

Magnetic survey can be used in foundation engineering to investigate the subsurface geology and to detect potential hazards that may affect the stability of the foundation. Magnetic survey involves measuring the variations in the earth's magnetic field to identify subsurface magnetic anomalies. These anomalies can provide information about the subsurface geology, including the presence of buried metallic objects, such as pipes, tanks or other underground structures.

Gravity survey is a geophysical method commonly used in geophysical exploration. It involves measuring the variations in the earth's gravitational field to identify subsurface density anomalies. Different rock types have different densities which can be detected using a gravimeter.

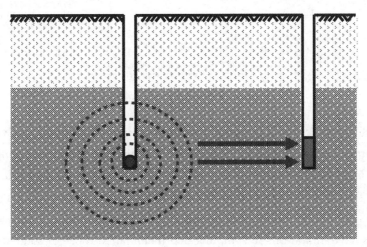

Figure 2-1   Schematic diagram of seismic reflection

2.2.2 Subsurface exploration and sampling

Subsurface exploration and sampling are methods used to investigate and analyze the properties and characteristics of the soil and rock beneath the earth's surface. This type of exploration is often conducted before construction projects or other activities that require

Chapter 2　Site Exploration and Characterization

knowledge of the underlying geology and soil conditions.

Subsurface exploration typically involves drilling boreholes or wells into the ground to collect samples of soil and rock. There are various types of drilling methods used for subsurface exploration, including auger drilling rotary drilling and percussion drilling, as shown in Figures 2-2 and 2-3. The specific method used will depend on the type of soil and rock being explored, as well as the depth of the exploration.

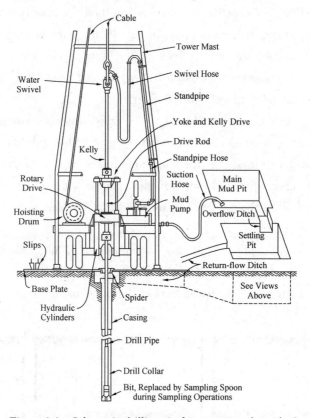

Figure 2-2　Schematic drilling rig for rotary wash methods

Disturbed sampling of soil provides a mean to evaluate stratigraphy by visual examination and to obtain soil specimens for laboratory index testing. Disturbed samples are usually collected using split-barrel samplers, as shown in Figures 2-4 and 2-5. Shallow disturbed samples can also be obtained by using hand augers and test pits. Direct push methods, such as GeoProbe sampling, can be used to obtain continuous disturbed samples. But these methods have limitations in sampling depth similar to those of solid stem and bucket augers. Samples obtained via disturbed sampling methods are generally used for index property testing in the laboratory such as density, moisture content, permeability and strength.

Through subsurface exploration, engineers can gather information on the properties of the soil and rock layers, such as their strength, stiffness and compressibility. This in-

(a) large diameter auger   (b) small diameter continuous flight auger

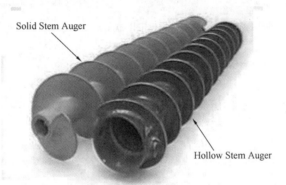

(c) solid stem auger and hollow stem auger

Figure 2-3 Various types of augers

formation is used to classify the soil and rock into different categories, which are then used to determine the appropriate foundation type.

For example, if the soil is found to be soft and compressible, such as in the case of clay or loose sand, a foundation system such as piles or deep foundations may be required to distribute the load of the structure over a larger area and prevent settlement. On the other hand, if the soil is found to be strong and stable, a shallow foundation system, such as a spread footing, may be sufficient.

In addition to soil and rock properties, subsurface exploration can also provide information on the presence of groundwater or other subsurface hazards that may affect foundation design. For example, if the groundwater level is high, a special foundation design may be necessary to prevent uplift and prevent the foundation from floating or heaving.

Chapter 2 Site Exploration and Characterization

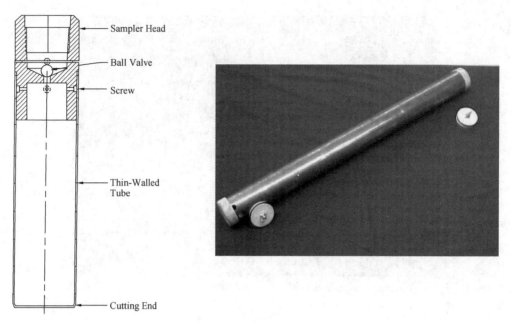

Figure 2-4 Schematic of thin-walled (shelby) tube and photo of tube with end caps

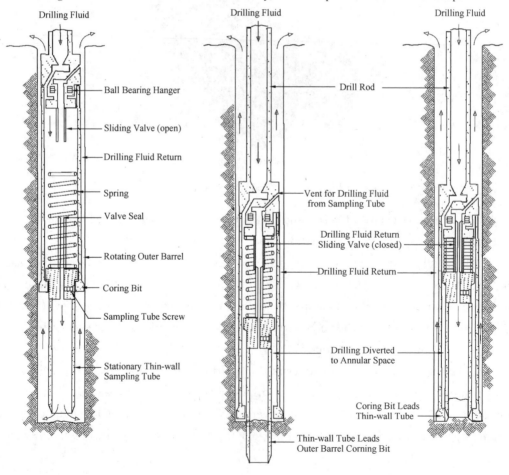

Figure 2-5 Pitcher sampler

2.2.3 Exploratory trenches

Exploratory trenches are typically dug using excavation equipments such as backhoes and bulldozers as shown in Figure 2-6. The trenches are typically between 1-2 m wide and can be several meters deep, depending on the requirements of the investigation.

Once the trench has been excavated, engineers and geologists inspect the soil and geology exposed by the trench walls. They take soil samples and conduct various tests to determine the soil's properties, such as its bearing capacity, shear strength and compressibility.

The information gathered from the exploratory trench investigation is used to determine the type of foundation that will be used for the building. For example, if the soil has a high bearing capacity and is relatively stable, a shallow foundation such as a spread footing or mat foundation may be appropriate. On the other hand, if the soil has a low bearing capacity or is unstable, a deep foundation such as a pile foundation may be necessary to support the weight of the building.

Exploratory trenches are an important tool in foundation engineering because they provide valuable information about the soil conditions and geology beneath the surface. This information is essential for designing a foundation that is both safe and structurally sound, and helps to ensure that the building will be stable.

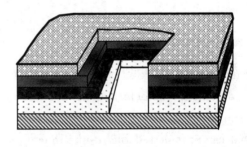

Figure 2-6 Exploratory trenches

2.3 In-situ testing

In-situ testing refers to testing or measurements that are performed on-site at the location where a particular material, structure or system is installed or in use. The term "in-situ" is derived from the Latin phrase meaning "in place". In foundation engineering, it is used to assess the properties of soil and rock formations at a construction site to design and optimize the foundation of a building, bridge or other structures.

Common types of in-situ tests in foundation engineering include: Plate Load Test (PLT), Cone Penetration Test (CPT), Standard Penetration Test (SPT), Dynamic Cone Penetration Test (DCPT), Pressure-meter Test (PMT), The Flat Dilatometer Test (DMT), and Vane Shear Test (VST), SPT, PMT and CPT, are shown in Figure 2-7. Several typical tests are further explained below.

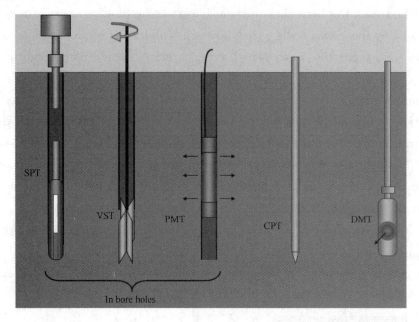

Figure 2-7 In-situ test methods

2.3.1 Plate load test (PLT)

A plate load test is a kind of in-situ test to simulate the load of the foundation, which is used to determine the deformation modulus of subgrade soil, the bearing capacity of subgrade soil and estimate the settlement of buildings. It is often considered as a test method that can provide reliable results in engineering. So this test is required when it is difficult to take undisturbed soil samples for important building foundations or complex foundations especially when it comes to loose sand or high-sensitivity soft clay.

During a plate load test, one or more hydraulic jacks are used to apply a known load to the foundation. The load is typically applied in increments, and the deformation and load response of the foundation are measured at each load increment. The deformation of the foundation is measured using displacement sensors, and the load response is measured using load cells. The data collected during the test is used to determine the load-deflection curve of the foundation, which provides information about the stiffness and capacity of the foundation. Figure 2-8 shows the layout of the plate load test and the load (p)-displacement (s) curve. The subgrade soil is considered to be failed and the test can be terminated until one of the following phenomena occurs.

(1) The soil around the load plate has obvious lateral extrusion or cracks.

(2) The load p increases slightly, but the settlement increases sharply, and the p-s curve has a steep drop.

(3) Under a certain load, the settlement rate cannot reach stable within 24 h.

If the above failure phenomenon does not occur, the foundation can still bear the load, but when the ratio of settlement to the width (or diameter) of the load plate $s/b(d)$ is $\geqslant$ 0.06, the test can also be terminated.

Draw a p-s curve according to the settlement produced by each level of load, as shown in Figure 2-8 (b). The front part of the curve is close to a straight line, which indicates that the subgrade soil has a linear deformation in Oa stage and no local plastic failure has occurred during this stage. The corresponding load is called plastic load or proportional limit load p_{cr}. The previous load before foundation failure is called ultimate load p_u.

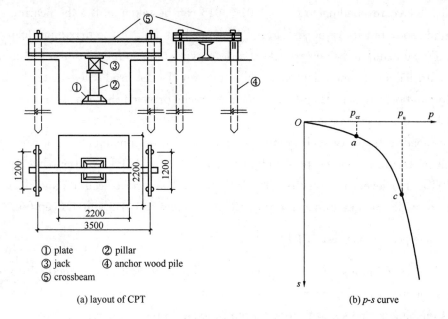

Figure 2-8 Plate Load Test (unit: mm)

The p-s curves can be used to obtain deformation modulus and the bearing capacity of the subgrade soil.

(1) Deformation modulus of subgrade soil

From the straight section of the p-s curve, the deformation modulus of soil E can be obtained by Equation 2-1:

$$E = \frac{pb(1-v^2)}{s}I \tag{2-1}$$

where

p —— the pressure corresponding to the settlement on the straight line section of the p-s curve, kPa;

b —— width of load plate, m;

v —— Poisson's ratio of soil, $v = 0.50$ for saturated soil;

I —— the coefficient reflecting the shape and stiffness of the load plate, and 0.886 for the rigid square load plate and 0.785 for the circular plate.

(2) Bearing capacity of subgrade soil

When determining the bearing capacity of subgrade soil by using the plate load test, the following criteria can be used according to the p-s curve.

① When the p-s curve has obvious straight line segments, the end points of the straight line segments p_{cr} can be taken as the bearing capacity of subgrade soil.

② when the ultimate load p_u can be determined from the p-s curve and p_u is less than two times of p_{cr}, take half p_u as the bearing capacity of subgrade soil.

③ When the above two criteria cannot be used, if the load plate area is 0.25-0.50 m², the load value corresponding to $s/b = 0.01$-0.015 can be taken as the the bearing capacity of subgrade soil, but its value should not be more than half of the maximum load.

Plate load tests can be performed on various types of deep foundations, including driven piles, drilled shafts and micropiles. The test can also be used to evaluate the integrity of the foundation, by detecting defects such as cracks or voids that may affect the capacity and behavior of the foundation.

Plate load tests are typically more expensive and time-consuming than other types of geotechnical testing, such as the standard penetration test (SPT) or the cone penetration test (CPT). However, the results of a plate load test are more accurate and reliable, and they can provide valuable information for the design and construction of deep foundations.

2.3.2 Cone penetration test (CPT)

Cone penetration test (CPT) is a widely used geotechnical testing method that is used to assess the mechanical properties and composition of soil and weak rock formations. It involves pushing a cone-shaped probe (often called a "cone penetrometer") into the ground at a constant rate and measuring the resistance encountered by the probe as it penetrates the soil, as shown in Figure 2-9.

The cone penetrometer is typically equipped with a series of sensors that measure various soil parameters, including tip resistance, sleeve friction and pore pressure, which is shown in Figure 2-10. These measurements are recorded continuously as the penetrometer is pushed into the ground, and are used to generate a detailed profile of the soil's mechanical properties.

Cone penetration testing is a versatile testing method that can be used in a wide range of applications, including foundation design, soil and groundwater contamination studies, and geotechnical engineering projects. It is particularly useful for investigating soft soils and unconsolidated sediments, where other testing methods may not be as effective.

2.3 In-situ testing

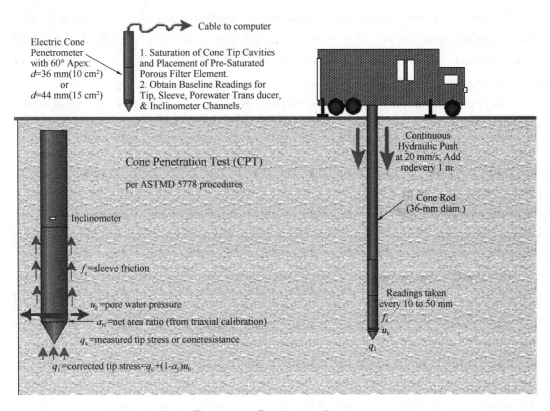

Figure 2-9   Cone penetration test

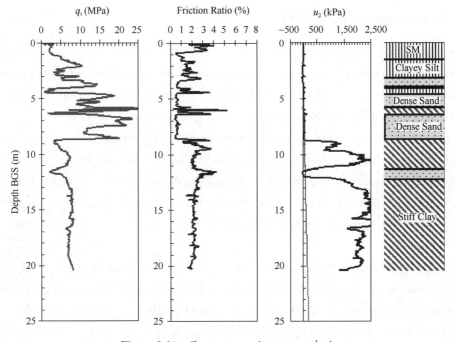

Figure 2-10   Cone penetration test analysis

The results of cone penetration testing can be used to determine soil strength stiffness, shear strength and other properties that are important for geotechnical design and construction. It is a widely accepted and reliable testing method that has been used for many years in a variety of applications.

The cone penetration test can be carried out from the ground surface with a need for a borehole. The test is carried out by first pushing the cone into the ground at a standard velocity of 1 to 2 cm/s while keeping the sleeve stationary.

For any depth, the resistance of the cone, called cone penetration resistance q_c, is recorded using the force probes provided for this purpose in the cone. Then the cone and the sleeve are moved and penetrated together into the soil. The combined cone and sleeve resistance, indicated by q_t, is recorded at any depth using tension load cells embedded in the sleeve.

This procedure is repeated and the measurements are made at regular depth intervals during penetration. In addition to the stress on the tip and the sleeve friction, the typical CPT probe measures as well the porewater pressure. Some equipped CPT probes are also able to measure shear wave velocity and temperature.

The cone penetration resistance values can be then correlated to shear strength parameters using proposed empirical curves. There are also some design methods associated with CPT results which directly use the CPT results to estimate the settlement in soils under a given pressure.

2.3.3 Standard penetration test (SPT)

The standard penetration test (SPT) is a widely used geotechnical testing method that measures the resistance of soil layers to penetration by a standard sampler driven by a standard weight dropped from a standard height, which is shown in Figure 2-11. It is a popular method used for site investigation, soil exploration and foundation design.

During an SPT, a borehole is drilled into the ground, and a standard sampler with a diameter of 2 in (50.8 mm) and a length of 18 in (457.2 mm) is inserted into the borehole. A standard weight of 140 lb (62.7 kg) is then dropped from a standard height of 30 in (762 mm) onto the top of the sampler, driving it into the soil. The number of blows required to drive the sampler the last 12 in (305 mm) of penetration is recorded as the SPT "N-value".

The SPT provides information about the relative density and strength of soil layers, particularly coarse-grained soils such as sands and gravels, which are shown in Table 2-1 and Figure 2-12. The results of the test can be used to estimate soil properties such as friction angle, cohesion and unit weight. The SPT is a relatively inexpensive and widely available testing method.

The SPT has some limitations and may not provide accurate results in certain soil conditions, such as very soft or very hard soils, or in soils with high moisture content. Therefore, it is important to use SPT results in conjunction with other geotechnical testing methods to obtain a complete understanding of soil properties and behavior.

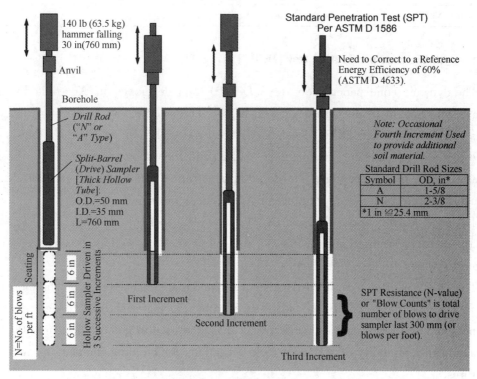

Figure 2-11　Standard penetration test

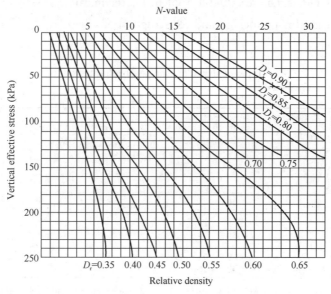

Figure 2-12　Relationship between relative density, N-value and vertical effective stress

Chapter 2   Site Exploration and Characterization

Sand density determined by SPT   Table 2-1

N value	Sand density	N value	Sand density
$N \leqslant 10$	Loose	$15 < N \leqslant 30$	Median dense
$10 < N \leqslant 15$	Slightly dense	$N > 30$	Dense

2.3.4   Dynamic cone penetration test (DCPT)

The dynamic cone penetration test (DCPT) was developed in Australia by Scala (1956). The current model was developed by the Transvaal Roads Department in South Africa (Luo, 1998). The DCPT consists of a steel rod with a cone-shaped tip that is connected to a sliding hammer. The hammer is lifted and then allowed to drop onto the rod, driving the cone into the ground. The depth of penetration is recorded for each blow of the hammer, and the resistance encountered by the cone is measured.

Table 2-2 shows various types of DCPT. The results of the DCPT are used to determine the density, strength and stiffness of the soil or rock, as well as its resistance to deformation, which are shown in Tables 2-3 and 2-4.

The DCPT is often used in road construction and maintenance projects where it can be used to evaluate the strength and stiffness of subgrade materials and to determine the thickness and uniformity of pavement layers. It is also used in geotechnical investigations to evaluate the suitability of soils and rock formations for building foundations and other structures.

Overall, the CDP test is a valuable tool in geotechnical engineering and can provide important information for the design and construction of a wide range of structures and infrastructure projects.

Types of dynamic cone penetration tests   Table 2-2

Type		Light	Heavy	Super-heavy
Drop hammer	Mass of hammer /kg	10	63.5	120
	Falling distance /cm	50	76	100
Probe	Diameter /mm	40	74	74
	Cone angle/(°)	60	60	60
Probe diameter /mm		25	42	50-60
Index		Penetration of 30 cm N_{10}	Penetration of 10 cm $N_{63.5}$	Penetration of 10 cm N_{120}
Suitable soil		Shallow fill, sand, silt and cohesive soil	Sandy soil, gravel soil below medium density and extremely soft rock	Dense gravel soil, soft rock and extremely soft rock

2.3 In-situ testing

Density classification of gravel soil by $N_{63.5}$　　　Table 2-3

$N_{63.5}$	Density	$N_{63.5}$	Density
$N_{63.5} \leqslant 5$	Loose	$10 < N_{63.5} \leqslant 20$	Median dense
$5 < N_{63.5} \leqslant 10$	Slightly dense	$N_{63.5} > 20$	Dense

Density classification of gravel soil by N_{120}　　　Table 2-4

N_{120}	Density	N_{120}	Density
$N_{120} \leqslant 3$	Loose	$11 < N_{120} \leqslant 14$	Dense
$3 < N_{120} \leqslant 6$	Slightly dense	$N_{120} > 14$	Very dense
$6 < N_{120} \leqslant 11$	Median dense		

2.3.5 Vane shear test (VST)

The vane shear test is an in-situ test for determining the undrained shear strength of soft to medium clays. Figure 2-13 is a schematic diagram of the essential components and test procedure. The test consists of forcing a four-bladed vane into undisturbed soil and rotating it until the soil shears. Two shear strengths are usually recorded, the peak shearing strength and the remolded shearing strength. These measurements are used to determine the sensitivity of clay, which is defined as the ratio of the peak undrained shearing strength to the remolded undrained shearing strength. Sensitivity, S_t, allows analysis of the soil resistance to be overcome during pile driving in clays which is useful for pile drivability analyses. It is necessary to measure skin friction along the steel connector rods which must be subtracted to determine the actual shear strength. The VST generally provides the most accurate undrained shear strength values for clays with undrained shear strengths less than 50 kPa.

It should be noted that the sensitivity of a clay determined from a vane shear test provides insight into the set-up potential of the clay deposit. However, the sensitivity value is a qualitative and not a quantitative indicator of soil set-up.

Example 2-1:

A plate load test was conducted on a construction site. The rigid plate is square with an area of 0.5 m². When the load is increased to 375 kPa, the soil around the plate has significant lateral extrusion. The measured test data are given in Table 2-5. Please determine the bearing capacity of the soil.

The plate load test data　　　Table 2-5

p(kPa)	25	50	75	100	125	150	175	200
s(mm)	0.80	1.60	2.41	3.20	4.00	4.80	5.60	6.40
p(kPa)	225	250	275	300	325	350	375	
s(mm)	7.85	9.80	12.1	16.4	21.5	26.6	43.5	

Chapter 2　Site Exploration and Characterization

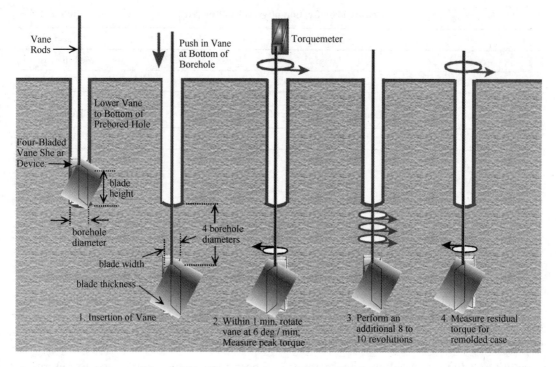

Figure 2-13　Vane shear test equipment and procedure

Solutions:

As shown in Table 2-5, the p-s curve has obvious straight line segments, the end points of the straight line segments $p_{cr}=200$ kPa.

When the load is increased to 375 kPa, the soil around the plate has significant lateral extrusion, $p_u = 350$ kPa;

The bearing capacity of the soil=min{200, 350/2}=175 kPa

2.4　In-situ monitoring

In foundation engineering, in-situ monitoring typically refers to the practice of measuring the performance of a foundation or soil structure in its natural location or environment. This type of monitoring is often used to evaluate the effectiveness of a foundation design, to verify that it is performing as expected, and to detect any issues or problems that may arise.

In-situ monitoring techniques in foundation engineering may include the use of instruments such as strain gauges, inclinometers, settlement plates, piezometers and vibration sensors to measure various aspects of the foundation such as deformation, settlement, load distribution and vibration. These instruments may be installed during construction or retrofitting of the foundation and left in place for ongoing monitoring.

In-situ monitoring in foundation engineering can provide valuable information about the behavior of a foundation or soil structure under different types of loads and environmental conditions, which can help engineers to refine their designs and improve the safety and durability of structures. Additionally, in-situ monitoring can provide early warning of potential issues or problems, allowing engineers to take corrective action before they become more serious.

2.4.1 Structure settlement measurement

Structure settlement measurement is the practice of measuring the vertical movement of a structure or building over time. Settlement refers to the gradual sinking or subsidence of a foundation or soil beneath a structure, which can lead to structural damage, safety concerns and other issues.

Structure settlement measurement is typically performed using settlement plates or other instruments that are installed beneath a building's foundation or at various locations throughout the structure. These instruments are designed to measure changes in elevation or vertical movement over time, typically using a reference point or benchmark that is fixed in place.

Settlement measurement is important for determining the stability and safety of a structure, as well as for evaluating the effectiveness of foundation designs and construction techniques. It is commonly used in building and construction projects to ensure that structures remain level and stable over time, and to identify potential settlement issues before they become more serious.

In addition to settlement plates, other techniques for measuring structure settlement may include inclinometers, laser scanning, and GPS monitoring, among others. These methods may be used alone or in combination to provide a more comprehensive picture of a structure's settlement behavior and help engineers to make informed decisions about foundation design and maintenance.

Structure settlement measurement involves several steps which are given as follows.

Planning: Before settling on a measurement technique, the engineer or technician must first assess the needs of the project and determine the most appropriate method of measurement based on factors such as the size and complexity of the structure, the type of soil or substrate, and the desired level of accuracy.

Instrument Installation: Once the measurement method has been chosen, the engineer or technician will install the necessary instruments or equipment at strategic locations within the structure or beneath its foundation. This may involve drilling holes or excavating the soil, depending on the type of instruments being used.

Data Collection: After the instruments have been installed, they will begin collecting data on the vertical movement or settlement of the structure over time. The frequency and duration of data collection will depend on the specific needs of the project and may involve periodic manual readings or automated monitoring systems.

Data Analysis: The data collected from the instruments will then be analyzed and interpreted to determine the rate and extent of settlement, as well as to identify any potential issues or problems that may need to be addressed.

Reporting: Finally, the engineer or technician will prepare a report summarizing the results of the settlement measurement, including any recommendations for further action or follow-up monitoring.

Throughout the process, it is important to ensure that the measurement instruments are properly calibrated, that the data collected is accurate and reliable, and that all issues or anomalies are promptly identified and addressed.

2.4.2 Excavation pit monitoring

Excavation pit monitoring is the practice of measuring and analyzing the stability of soil and rock structures surrounding an excavation site or construction pit. This type of monitoring is critical for ensuring the safety of workers, nearby structures and the environment during excavation and construction activities.

Excavation pit monitoring typically involves the use of various instruments and techniques to measure and track factors such as ground movement, soil and rock stability, water levels, and other environmental conditions. Common techniques used in excavation pit monitoring include the following items.

Ground movement monitoring: Ground movement monitoring involves measuring the distance and direction of movement of the soil and rock structures surrounding the excavation pit. This can be done using surveying instruments such as total stations or GPS receivers, which are set up at fixed points around the excavation pit and used to measure changes in position over time. Ground movement monitoring is important for identifying potential instability or deformation of the ground and can help to prevent collapses or other hazards.

Inclinometers: Inclinometers are instruments that are used to measure the angle of soil or rock layers. They consist of a probe that is inserted into a borehole drilled into the ground, and they can be used to track any changes in the slope or angle of the soil or rock layers surrounding the excavation pit. Inclinometers are particularly useful for detecting potential instability in areas where there are steep slopes or where the soil or rock layers are prone to sliding or deformation.

Settlement monitoring: Settlement monitoring involves measuring the vertical movement of the ground surface or nearby structures during and after the excavation process. This can be done using surveying instruments such as total stations or leveling devices.

Crack monitoring: Crack monitoring involves tracking the development and movement of cracks in nearby structures that may be caused by the excavation process.

By monitoring the stability of the soil and rock structures surrounding a foundation excavation, engineers can take necessary precautions and implement mitigation measures to prevent potential hazards such as landslides, slope failures and. structural collapse. Excavation pit monitoring in foundation engineering can help to ensure that foundation excavation and construction activities are carried out safely and effectively with minimal impact on the surrounding environment and infrastructure.

2.4.3 Groundwater monitoring

Groundwater monitoring is an important aspect of foundation engineering because groundwater can have a significant impact on the stability and performance of building foundations. Here are some additional details on how groundwater monitoring works and why it's important.

Groundwater monitoring involves measuring the water levels and flow around a building foundation or excavation site. This is typically done by installing wells or piezometers (which are devices that measure groundwater pressure) at various depths around the site.

The data collected from the wells or piezometers is then used to determine the water level fluctuations over time. This information can help engineers assess the risk of water infiltration into the foundation, as well as the potential for soil instability or erosion.

Groundwater monitoring can help engineers identify potential issues before they become major problems. For example, if the data shows that groundwater levels are rising, engineers can take steps to prevent water infiltration into the foundation or implement drainage systems to manage the water.

In addition to measuring groundwater levels, groundwater monitoring can also involve analyzing the water chemistry. Changes in water chemistry can indicate potential issues with soil stability or erosion, which can affect the performance of the foundation.

Groundwater monitoring is often required by local building codes or regulatory agencies to ensure that building foundations are safe and stable. It can also be a proactive measure taken by property owners or developers to prevent future problems.

Overall, groundwater monitoring is an important tool in foundation engineering that helps ensure the safety and stability of building foundations. By monitoring groundwater levels and flow, engineers can identify potential issues and implement measures to prevent problems before they occur.

2.5 Main contents of the geotechnical investigation report

A geotechnical investigation report is a document that provides a detailed assessment of the soil, rock and groundwater conditions at a specific site. The report is typically prepared by a geotechnical engineer and is used to inform the design and construction of a structure. The main contents of a geotechnical investigation report include the following parts.

Introduction: This section provides background information on the project, including its purpose, scope and location.

Site Description: This section provides a description of the site, including its topography, geology and hydrology.

Field Investigation: This section describes the field investigation techniques used to collect soil and rock samples, as well as groundwater and other data.

Laboratory Testing: This section describes the laboratory tests performed on the soil and rock samples collected during the field investigation. These tests may include physical, mechanical and chemical analyses.

Geotechnical Analysis: This section presents the geotechnical analysis of the site, including an assessment of the soil and rock properties, groundwater conditions and potential geotechnical hazards.

Foundation Design Recommendations: This section provides recommendations for the design and construction of foundations based on the geotechnical analysis. These recommendations may include the type of foundation to be used, the depth and size of the foundation, and any special design considerations.

Slope Stability Analysis: If the site includes slopes or embankments, this section will include an analysis of slope stability and recommendations for slope stabilization measures.

Seismic Analysis: If the site is located in a seismic zone, this section will include an analysis of seismic hazards and recommendations for seismic design.

Conclusions and Recommendations: This section summarizes the findings of the geotechnical investigation and provides recommendations for the design and construction of the project.

Appendices: This section includes supporting documents and data, such as field and laboratory test results, maps and photographs.

Questions

2-1 What should be done for site exploration?

2-2 What is geological exploration? What methods are commonly used?

2-3 Standard penetration test (SPT) is often used to determine the compact degree of sand. Try to explain the method used to determine the compact degree of sand.

2-4 What is cone penetration test (CPT)? What are the characteristics of single-bridge CPT and double-bridge CPT?

2-5 Plate load test is a very important in-situ test. What main results can be obtained?

2-6 In a plate load test, what is the proportional limit load p_{cr} and what is the ultimate load p_u? How to determine the bearing capacity of subgrade?

2-7 What is the vane shear test (VST)? Which kind of soil is it often used to determine its shear strength? What kind of shear strength can be obtained by VST?

2-8 The plate load test is conducted on the gravel layer. The plate area $A=0.71\text{m}\times 0.71\text{m}$. The pressure-settlement (p-s) curve obtained is shown in Figure 2-14, and no obvious straight line segment can be found in p-s curve. Try to estimate the deformation modulus E of the gravel layer.

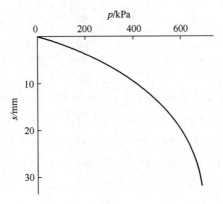

Figure 2-14 The p-s curve obtained in plate load test

Chapter 3   Design of Shallow Foundation on Natural Subgrade

3.1   Introduction

Foundations significantly influence the safety and cost of structures, which affects the design and construction of structures. To design reasonable and economical foundations, two factors must be considered. The first one is the properties of the structure which includes its function, safety grade, layout, loadings, etc. The second one is the engineering geology and hydrogeology condition of the subgrade which includes the soil layer distribution, mechanical properties of the soil, groundwater level, etc..

The subgrade can be divided intothe natural subgrade and artificial subgrade. The natural soil layer where the foundation is placed directly is called the natural subgrade. If the natural subgrade is too weak or has poor engineering geological properties, it requires artificial reinforcement or treatment before a foundation can be built. This is known as an artificial subgrade. The depth of embedment of the foundation on the natural subgrade significantly affects the foundation's construction methods, foundation type and design method. The foundation can be generally divided into shallow foundation and deep foundation according to its depth of embedment. Usually, the foundation with an embedding depth generally less than 5 m is called a shallow foundation. The side friction between the foundation and the surrounding soil can be ignored in the design and calculation of the shallow foundation. The shallow foundation has the advantage of convenient construction, simple technology and economic cost so that it is the most commonly used foundation type for structures. This chapter focuses on the design of shallow foundations on the natural subgrade.

3.1.1   Design grade of foundation

The safety and normal use of structures depends not only on the safety reserve of the superstructure but also on the safety degree of the foundation. Because foundations are underground projects, even small damages in the foundation can be difficult to repair. Therefore, foundation design plays an essential role in the structure design. The foundation design should consider the complexity of the subgrade, the size, function, and characteristics of the structure, as well as the potential damage or impact that the subgrade can have on the structure. Chinese foundation design code named *Code for Design of Building Foundation* (GB 50007—2011) divides the design of the foundation into three grades, which are detailed in Table 3-1.

Design Grade of Foundation Table 3-1

Design grade	Type of subgrade and foundation
Grade A	Important industrial and civil structures. High-rise structures with more than thirty floors. Complex structures with top and bottom floors having a height difference of more than ten floors connected as one integrated structure. Large areas with multi-floored underground structures (such as underground garages, shopping malls, sports grounds, etc.). Structures have special requirements for subgrade deformation. Structures on slopes (including high slopes) under complex geological conditions. New Structures significantly affect the original projects. General structures with the complex site and subgrade conditions. Excavation pit with no less than two floors of a basement in complex geological conditions and soft soil areas
Grade B	Industrial and civil structures except for Grade A and Grade C
Grade C	No more than seven floors of civil structures and general industrial structures with simple site conditions and uniform load distribution; secondary light structures. Excavation pits in non-soft soil areas with simple site geological conditions, simple environmental conditions, low environmental protection requirements and excavation depth of less than 5.0 m

3.1.2 Design principle of foundation

The purpose of foundation engineering design is to create a safe, economical and feasible foundation that ensures the safety and proper functioning of the superstructure. Therefore, the basic principles for designing and calculating foundation engineering can be outlined as follows:

(1) The subgrade should have sufficient strength to meet the requirements of subgrade bearing capacity.

(2) The deformation of the subgrade and foundation can meet the allowable requirements of the normal use of structures.

(3) The overall stability of the subgrade and foundation is sufficiently guaranteed.

(4) The foundation itself has sufficient strength, rigidity and durability.

The type of subgrade and foundation depends primarily on the engineering geological and hydrogeological conditions of the subgrade soil layer, the type of superstructure, the load conditions, the normal usage requirements, the construction materials and technology, etc. Different types of foundations should be compared, and a more suitable and reasonable foundation should be selected.

According to the Chinese *Code for Design of Building Foundation* (GB 50007—2011), foundation design must satisfy the requirements of both bearing capacity limit state and normal service limit state. The foundation design needs to meet specific standards based on the design grade and the extent to which foundation deformation affects the superstructure under long-term loads. Below are the requirements.

(1) The subgrade of all grades of structures shall meet the bearing capacity requirements.

(2) Structures with Grades A and B should meet the subgrade deformation requirements.

(3) For structures with Grade C, those listed in Table 3-2 may not be checked for deformation unless otherwise specified.

(4) High-rise structures, retaining walls, and structures built on or near slopes are often subjected to horizontal loads. Therefore it is important to assess their stability.

(5) The stability of the excavation pit should be assessed.

(6) When the underground water is shallow, and there is a floating problem for the basement or underground structures, an anti-floating evaluation should be carried out.

Grade C structures without the requirement of subgrade deformation   Table 3-2

Main bearing soil layers of subgrade	Characteristic value of subgrade bearing capacity f_{ak}/kPa		$80 \leqslant f_{ak} \leqslant 100$	$100 \leqslant f_{ak} \leqslant 130$	$130 \leqslant f_{ak} \leqslant 160$	$160 \leqslant f_{ak} \leqslant 200$	$200 \leqslant f_{ak} \leqslant 300$
	Slope of soil layer/%		$\leqslant 5$	$\leqslant 10$	$\leqslant 10$	$\leqslant 10$	$\leqslant 10$
Type of structures	Layer number of masonry structures and frame structures		$\leqslant 5$	$\leqslant 5$	$\leqslant 6$	$\leqslant 6$	$\leqslant 7$
	Single-layer bent structure (6 m column spacing)	single-span — Rated lifting weight of crane /t	10-15	15-20	20-30	30-50	50-100
		single-span — Workshop span /m	$\leqslant 18$	$\leqslant 24$	$\leqslant 30$	$\leqslant 30$	$\leqslant 30$
		Multi-span — Rated lifting weight of crane /t	5-10	10-15	15-20	20-30	30-75
		Multi-span — Workshop span /m	$\leqslant 18$	$\leqslant 24$	$\leqslant 30$	$\leqslant 30$	$\leqslant 30$
	Chimney	Height /m	$\leqslant 40$	$\leqslant 50$	$\leqslant 75$		$\leqslant 100$
	Water tower	Height /m	$\leqslant 20$	$\leqslant 30$	$\leqslant 30$		$\leqslant 30$
		Volume /m³	50-100	100-200	200-300	300-500	500-1000

Note: 1. The main bearing soil layer of a subgrade refers to the depth of 3 b (where b is the width of the foundation's bottom surface) under the bottom surface of a strip foundation, and the depth of 1.5 b with a minimum thickness of 5 m under the bottom surface of a pad foundation, except for ordinary civil buildings with floors less than two.

       2. Masonry structures and frame structures in the table refer to civil buildings, and industrial buildings can be converted into equivalent civil building floors according to the height and load of the factory building.

3.1.3 Design loads

The foundation may be subjected to various types of loads, including permanent, live, accidental and seismic loads. Therefore, it is important to consider the combined action of these loads and use the most unfavourable load combination in the design. According to the Chinese *Code for Design of Building Foundation* (*GB 50007—2011*), the most unfavourable load combination and corresponding resistance limit should be used in the design of the foundation as follows.

(1) When determining the bottom area and buried depth of the foundation based on the foundation's bearing capacity, or when determining the number of piles based on the bearing capacity of a single pile, it is important to combine the loads transmitted to the bottom of the foundation or pile cap according to the standard load combination under the normal service limit state. The corresponding resistance should be based on the characteristic value of the subgrade bearing capacity or the bearing capacity of a single pile.

(2) When calculating subgrade deformation, it is important to consider the loads transmitted to the bottom of the foundation using the quasi-permanent load combination under the normal service limit state. However, wind and earthquake loads should not be included in this calculation. The corresponding limit value should be based on the allowable value of subgrade deformation.

(3) When calculating the earth pressure of a retaining wall, the stability of a subgrade or slope, or landslide thrust, the loads should be based on the fundamental load combination under the bearing capacity limit state. In this case, the partial factors should all be 1.0.

(4) When determining the height of a foundation or pile abutment, the cross-section of a retaining structure, the internal force of a foundation or retaining structure, and when checking the strength of materials and reinforcement, it is important to use the corresponding partial coefficients according to the fundamental load combination under the bearing capacity limit state. Additionally, when it is necessary to check the crack width of the foundation, the standard load combination should be used according to the normal service limit state.

(5) The design safety grade of the foundation, the service life of the structure, and the structural importance coefficient should be determined according to relevant codes. However, the structural importance coefficient should not be less than 1.0.

3.1.4 Steps in the design of the shallow foundation

The design of shallow foundation on natural foundation usually follows the following steps.

(1) Select the material and type of foundation and determine the layout.

(2) Choose the buried depth of the foundation, that is, determine the bearing soil lay-

er of the foundation.

(3) Determine the characteristic value of subgrade bearing capacity.

(4) Determine the bottom area of foundation according to the load effect transmitted to the bottom of foundation and the characteristic value of subgrade bearing capacity.

(5) Evaluate the subgrade deformation and stability according to the loads transmitted to the bottom of the foundation.

(6) Determine the structure size of the foundation and make necessary structural calculations according to the loads transmitted to the bottom of the foundation.

(7) Finish foundation working drawings.

3.2 Types of the shallow foundation

A shallow foundation on natural subgrade can be divided into pad foundation, strip foundation (including cross strip foundation), raft foundation, box foundation, shell foundation, etc., according to the structure type of foundation.

(1) Pad foundation

Pad foundation is generally used as the foundation of a column (Figure 3-1).

(2) Strip foundation

This is the foundation of a wall that is usually set in a continuous strip shape (Figure 3-2).

If the load of the column is large and the bearing capacity of the soil layer is low, a large area pad foundation is usually needed. In such a case, the strip foundation (Figure 3-3) or cross-beam foundation (Figure 3-4) is a better option.

If the building is relatively light and the load acting on the wall is not substantial, and the foundation needs to be built on a deep, stable soil layer, constructing a strip foundation may not be cost-effective. Instead, a lintel can be added under the wall and supported on a pad foundation. This type of foundation is commonly referred to as a pad foundation under the wall, as shown in Figure 3-5.

Figure 3-1　Pad foundation under column

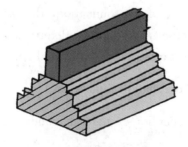

Figure 3-2　Strip foundation under wall

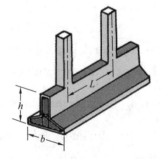

Figure 3-3　Strip foundation under column

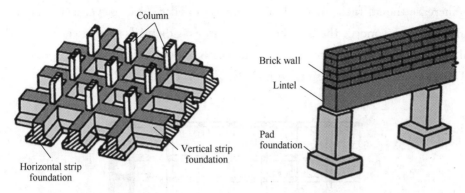

Figure 3-4　Cross-beam foundation under column

Figure 3-5　Pad foundation under wall

(3) Raft foundation and box foundation

If the load from the column or wall is significant, and the subgrade soil is weak, resulting in insufficient subgrade bearing capacity for a pad foundation or strip foundation, or if the groundwater level remains above the basement floor throughout the year, it may be necessary to use a raft foundation to prevent groundwater from infiltrating into the room. A raft foundation involves making the entire bottom of the house (or basement section) into a continuous reinforced concrete slab, as shown in Figure 3-6.

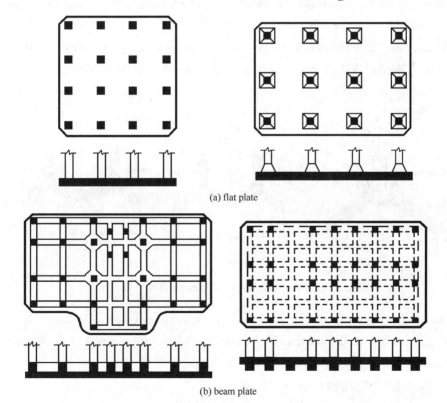

(a) flat plate

(b) beam plate

Figure 3-6　Raft foundation in basement

To increase the stiffness of the foundation slab and reduce uneven settlement, high-rise buildings often connect the basement floor, roof, side walls, and a certain number of internal partitions to create a reinforced concrete box structure with strong overall stiffness. This type of foundation is called box foundation (Figure 3-7).

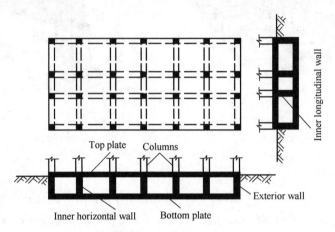

Figure 3-7 Box foundation

(4) Shell foundation

To enhance the mechanical performance of the foundation, the shape of the foundation can be designed as various types of shells instead of steps. This type of foundation is known as a shell foundation (Figure 3-8).

Tall structures like chimneys, water towers, and TV towers are often built with shell foundations. The arch effect can be utilized to make the internal forces of the structure more reasonable.

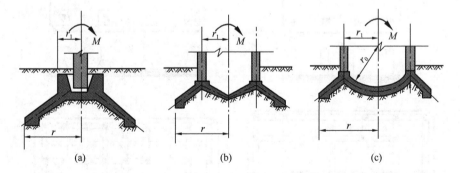

Figure 3-8 Various structural form of shell foundation

Shallow foundations can be divided into two types: unreinforced (rigid) spread footings and reinforced spread footings. The difference between the two lies in the presence or absence of reinforcement within the spread footing.

(1) Unreinforced (rigid) spread footings

Unreinforced (rigid) spread footings are typically made of materials such as brick, block stone, rubble, plain concrete, and lime soil, without any steel reinforcement. Although these materials possess good compressive properties, their tensile and shear strengths are low. In design, it is necessary to limit the ratio of the extended width to the height of the foundation to prevent the tensile and shear stresses in the foundation from exceeding its material strength. As a result, the relative height of the foundation is generally high, and little bending deformation occurs. This type of foundation is commonly referred to as a rigid foundation and its cross-section is shown in Figure 3-9. Rigid foundations are suitable for civil buildings with up to six floors (three-story concrete foundations should not exceed four floors) and masonry-bearing factory buildings. They are characterized by their good stability, simple construction, and ability to bear large loads.

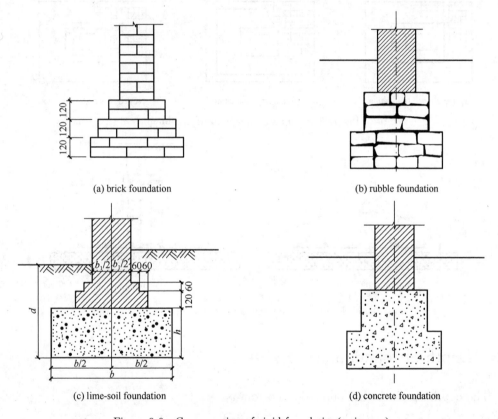

Figure 3-9 Cross-section of rigid foundation (unit: mm)

(2) Spread footings

When the foundation bears a large external load including bending moments and horizontal loads, and the bearing capacity of the subgrade is low, the above rigid foundation cannot meet the requirements of the bearing capacity and burial depth of the foundation, reinforced concrete foundation can be considered. Reinforced concrete foundations (spread footings) can meet the requirements of subgrade bearing capacity by increasing the bottom area of the foundation without increasing its burial depth. Common spread footings include pad foundations under columns, strip foundations, cross shaped strip foundations, raft and box foundations, which have good performance to resist bending and shear. Figures 3-10 and 3-11 show the pad foundation under column and reinforced concrete strip foundation under the wall respectively.

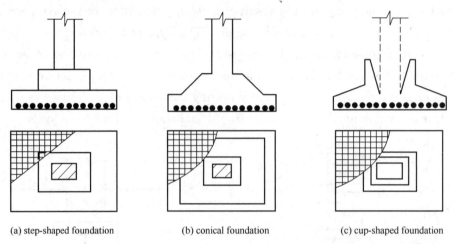

(a) step-shaped foundation　　　　(b) conical foundation　　　　(c) cup-shaped foundation

Figure 3-10　Pad foundation under column

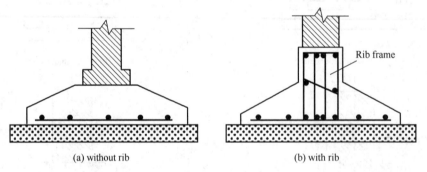

(a) without rib　　　　(b) with rib

Figure 3-11　Reinforced concrete strip foundation under the wall

3.3　Depth of foundations

The depth at which the foundation bottom is buried below the ground is known as the buried depth of the foundation. To ensure the safety of the foundation and reduce its size, it is preferable to place the foundation on a good soil layer. However, if the foundation is buried too deep, it can be inconvenient to construct and increase the cost of the foundation. Therefore, a reasonable buried depth should be chosen based on the specific circumstances. Figure 3-12 shows the minimum em-

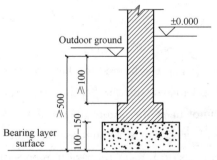

Figure 3-12　Minimum embedded depth of foundation (unit: mm)

bedded depth of foundation. For all foundation types except rock foundations, the buried depth should not be less than 0.5 m. Several factors influence the buried depth of a foundation, with the most important being the following four aspects.

3.3.1 Function, type and loads of buildings

The function and purpose of a building are often the primary factors in determining the appropriate buried depth of the foundation. Buildings equipped with basements, underground facilities, or semi-buried structures require a greater foundation burial depth. When there is a basement, the buried depth of the foundation will be affected by the ground elevation of the basement. In cases where there is only a local basement on the plane, the buried depth of the foundation can be modified with steps or deepened as a whole. When determining the buried depth of the foundation, it is important to consider the elevation of water supply and drainage, heating, and other pipelines. In general, pipelines should not pass under the foundation. Instead, a hole can be created in the foundation with sufficient clearance between the top surface of the hole and the pipeline to prevent foundation settlement from causing pipeline fractures and accidents. When determining the buried depth of the foundation for a cold storage or high-temperature furnace, the potential adverse effects of heat conduction on the foundation soil should be considered, such as frost heaving due to low temperature or drying shrinkage due to high temperature.

High-rise buildings with significant vertical and horizontal loads, such as earthquake and wind forces, require an appropriate increase in the buried depth of the foundation to ensure stability. In areas with seismic fortification, the embedded depth of box and raft foundations on natural foundation, excluding rock foundation, should not be less than 1/15 of the building height. The embedded depth of pile box or raft foundation (excluding pile length) should not be less than 1/20 to 1/8 of the building height. High-rise buildings on rock foundation should meet anti-sliding requirements. For structures subjected to uplift force, such as transmission towers, the buried depth of the foundation should be sufficient to meet uplift requirements. For workshops and warehouses with large indoor ground loads or equipment foundations, any adverse effects on the foundation's interior should be taken into account.

Simply supported beam bridges with medium and small spans have little impact on determining the buried depth of the foundation. However, for statically indeterminate structures, even a slight uneven displacement of the foundation can cause a significant change in internal force, such as the abutment of an arch bridge. To minimize potential horizontal displacement and settlement differences, the foundation may need to be placed on a deeply buried, solid soil layer.

3.3.2 Engineering geological and hydrogeological conditions

When determining the buried depth of shallow foundations, it is important to carefully analyse geological exploration data and try to place the foundation on good soil whenever possible. However, the quality of soil is relative. The same soil layer may meet the bearing capacity requirements for light buildings and serve as a suitable natural foundation, but it may not meet the bearing capacity requirements for heavier structures and may not be appropriate as a natural foundation. Therefore, when considering foundation factors, it is necessary to take into account the nature of the building. Different geogenic soil layers have unique properties and can be broadly classified into the following five typical situations.

(1) In the first case (Figure 3-13a), the interior of the foundation consists entirely of good soil (with high bearing capacity, uniform distribution, and small compressibility), and the buried depth of the foundation is primarily determined by other factors.

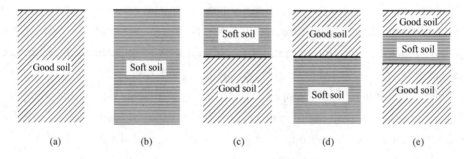

Figure 3-13 Composition types of foundation soil layer

(2) In the second case (Figure 3-13b), the foundation is composed of soft soil with high compressibility and limited bearing capacity. In such cases, shallow foundations on natural soil are generally not appropriate. For low-rise buildings, if shallow foundations are used, corresponding measures should be taken, such as enhancing building stiffness.

(3) In the third case (Figure 3-13c), the foundation consists of two soil layers, with a soft upper layer and a good lower layer. The buried depth of the foundation should be determined based on the thickness of the soft soil layer and the type of building. This situation can be further divided into the following three scenarios.

① When the thickness of the soft soil layer is less than 2 m, the foundation should be constructed on the good soil at the lower level.

② If the soft soil layer is 3-4 m thick, for low-rise buildings, the foundation may be constructed in the soft soil layer to avoid extensive excavation. However, the superstructure's rigidity should be appropriately enhanced. For important buildings and buildings with basements, the foundation should be built on the good soil layer beneath the soft soil.

③ If the thickness of the soft soil layer is more than 5 m, in addition to using large-scale foundations like rafts or boxes, and for situations similar to the second case mentioned earlier, foundation treatment or pile foundations may also be necessary.

(4) In the fourth case (Figure 3-13d), the foundation consists of two soil layers, with a good upper layer and a soft lower layer. In such cases, the foundation should be buried as shallow as possible to minimize pressure on the soft soil layer, and the bearing capacity of the underlying soft layer should be checked. If the good soil layer is very thin, it should be considered similar to the second case mentioned earlier.

(5) In the fifth case (Figure 3-13e), the foundation consists of alternating layers of good and soft soil. In this scenario, the buried depth of the foundation should be selected based on the thickness and bearing capacity of each soil layer, and with reference to the principles mentioned above.

To prevent drainage or precipitation in the foundation pit during construction, the foundation should be buried above the groundwater level as much as possible. If there is confined water in the foundation, it's necessary to check whether excavating the foundation trench will damage the basement's water-resistant layer above the confined water layer, due to the flow of soil caused by the floating action of pressure water.

3.3.3 Influence of adjacent buildings

In densely built urban areas, new buildings' foundation depths should not be greater than those of the original buildings to ensure safety and normal usage of existing structures. Additionally, the adverse effects of new loads on existing structures should be considered. When a new building has a heavy load and a deep foundation that exceeds the original building's foundation depth, a certain clear distance from the original foundation should be considered during the design (Figure 3-14). This distance should be determined based on the original building's load, soil condition, and foundation type. Typically, it should be at least 1-2 times the height difference of adjacent foundation bottoms, i.e., $L \geqslant (1-2)\Delta H$. When clear distance requirements cannot be met, segmented construction or temporary support, sheet piling, underground continuous wall and other measures should be implemented. Alternatively, the original building's foundation should be reinforced.

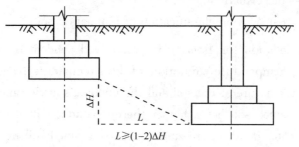

Figure 3-14 Effects of adjacent foundations

3.3.4 Frost heave of subgrade

Within a certain depth of the ground, the temperature of the soil changes with the seasons. In cold regions during winter, the water in the topsoil freezes due to decreasing temperatures. When the water freezes, the volume of water in the soil expands, causing the whole soil layer's volume to expand. However, this volume expansion is limited and the frozen soil generates suction, which attracts nearby water to seep into the freezing area and freeze. As a result, the soil experiences an increase in water content, and the volume expands, causing the phenomenon known as frost heaving. When the frozen soil layer is close to the groundwater level, the suction and capillary force generated by freezing attract groundwater to enter the frozen soil area to form ice crystals. In severe cases, ice interlayers are formed, and the ground swells due to the frost heave of the soil. In spring, when the temperature rises, the frozen soil layer shrinks in volume, and it also melts due to the significant increase in water content and the significant decrease in strength. Frost heave and thaw settlement are uneven, and if there is a thick frozen soil layer under the basement, it may produce incalculable frost heave and thaw settlement deformation.

The frost heaving of soil is influenced by the soil properties and the freezing conditions of the surrounding environment, as well as the availability of water in the soil area. Soil with coarser particles and strong water permeability will allow unfrozen water to be discharged from the frozen area during freezing, resulting in smaller frost heaving. High-plasticity clay, on the other hand, with a plasticity index greater than 22 and low water permeability, is difficult to get water supplement during freezing and has low frost heaving properties. Higher natural water content, especially free water content, also leads to stronger frost heaving. The closer the frozen soil area is to the groundwater level, the greater the frost heaving. The frost heaviness of soil can be measured by the average frost heaving rate η:

$$\eta = \frac{\Delta z}{h' - \Delta z} \times 100\% \qquad (3\text{-}1)$$

where

Δz —— frost heaving of the surface;

h' —— thickness of permafrost.

Table 3-3 gives frost heaving classification of foundation soil.

The depth of foundation soil freezing is primarily dependent on local meteorological conditions, including temperature, duration of low temperature and intensity of cold weather. Secondly, the nature of the soil and the building's environmental conditions also play a role. Coarse-grained soil has a higher thermal conductivity than fine-grained soil, which can cause a greater depth of freezing for coarse-grained soil under the same conditions. The water content of the soil also affects the depth of freezing, as water releases la-

3.3 Depth of foundations

tent heat when it freezes. The greater the water content, the shallower the freezing depth. Additionally, in dense urban areas, high-rise buildings can absorb heat from the outside and contribute to the "heat island effect", where temperatures are higher than those of the surrounding areas, leading to a shallower freezing depth. Industrial facilities, transportation, winter heating, and human activity also contribute to the heat island effect.

Frost heaving classification of foundation soil Table 3-3

Name of soil	Natural moisture content before freezing w (%)	The minimum distance between groundwater level and freezing surface during freezing h_w (m)	Average frost heaving rate η (%)	Frost heave grade	Frost heave category
Crushed (ovoid) stone, gravel, coarse sand, medium sand (particle size less than 0.075 mm with particle content greater than 15%) and fine sand (particle size less than 0.075 mm with particle content greater than 10%)	$w \leqslant 12$	>1.0	$\eta \leqslant 1$	I	Non-frost heave
	$12<w \leqslant 18$	$\leqslant 1.0$	$1<\eta \leqslant 3.5$	II	Weak frost heaving
		>1.0			
	$w>18$	$\leqslant 1.0$	$3.5<\eta \leqslant 6$	III	Frost heaving
		>0.5			
		$\leqslant 0.5$	$6<\eta \leqslant 12$	IV	Strong frost heaving
silt	$w \leqslant 14$	>1.0	$\eta \leqslant 1$	I	Non-frost heave
		$\leqslant 1.0$	$1<\eta \leqslant 3.5$	II	Weak frost heaving
	$14<w \leqslant 19$	>1.0			
		$\leqslant 1.0$	$3.5<\eta \leqslant 6$	III	Frost heaving
	$19<w \leqslant 23$	>1.0			
		$\leqslant 1.0$	$6<\eta \leqslant 12$	IV	Strong frost heaving
	$w>23$	Not considered	$\eta>12$	V	Extra strong frost heaving
	$w \leqslant 19$	>1.5	$\eta \leqslant 1$	I	Non-frost heave
		$\leqslant 1.5$	$1<\eta \leqslant 3.5$	II	Weak frost heaving
	$19<w \leqslant 22$	>1.5	$1<\eta \leqslant 3.5$	II	Weak frost heaving
		$\leqslant 1.5$	$3.5<\eta \leqslant 6$	III	Frost heaving
	$22<w \leqslant 26$	>1.5			
		$\leqslant 1.5$	$6<\eta \leqslant 12$	IV	Strong frost heaving
	$26<w \leqslant 30$	>1.5			
		$\leqslant 1.5$	$\eta>12$	V	Extra strong frost heaving
	$w>30$	Not considered			

41

continued

Name of soil	Natural moisture content before freezing w (%)	The minimum distance between groundwater level and freezing surface during freezing h_w (m)	Average frost heaving rate η (%)	Frost heave grade	Frost heave category
cohesive soil	$w \leqslant w_P + 2$	>2.0	$\eta \leqslant 1$	I	Non-frost heave
		≤2.0	$1 < \eta \leqslant 3.5$	II	Weak frost heaving
	$w_P + 2 < w \leqslant w_P + 5$	>2.0			
		≤2.0	$3.5 < \eta \leqslant 6$	III	Frost heaving
	$w_P + 5 < w \leqslant w_P + 9$	>2.0			
		≤2.0	$6 < \eta \leqslant 12$	IV	Strong frost heaving
	$w_P + 9 < w \leqslant w_P + 15$	>2.0			
		≤2.0	$\eta > 12$	V	Extra strong frost heaving
	$w > w_P + 15$	Not considered			

Note: 1. w_P ——percentage plastic limit water content;
   w ——percentage average value of natural water content before freezing in permafrost.
2. Salinized frozen soil is not listed in the table.
3. When the plasticity index is greater than 22, the frost-heaving property decreases by one level.
4. When the content of particles with a particle size less than 0.005 mm is more than 60%, it is non-frost heave soil.
5. The frost-heaving properties of gravel soil are determined based on the type of filling material when it constitutes more than 40% of the total mass.
6. Gravel soil, gravel sand, coarse sand, medium sand (with a particle size less than 0.075 mm and a particle content less than 15%), and fine sand (with a particle size less than 0.075 mm and a particle content less than 10%) are all considered non-frost heave soil.

Soil that remains frozen for two years or more is referred to as permafrost, while soil that freezes and melts periodically with seasonal changes is known as seasonal frozen soil. When designing a foundation on seasonal frozen soil, the freezing depth of the site must be calculated using the following equation:

$$z_d = z_0 \, \psi_{zs} \, \psi_{zw} \, \psi_{ze} \tag{3-2}$$

where

z_0 ——standard freezing depth, for non-frost heaving cohesive soil, the average value of the maximum freezing depth measured in the open field outside the city with a flat and bare surface for not less than ten years is adopted. When there is no measured data, the value can be taken according to the standard freezing depth map given in the *Code for the Design of Building Foundation* (*GB 50007—2011*) in China. For example, the z_0 values of some major cities in northern China are as follows:

Jinan   0.5 m   Xi'an   0.5 m   Tianjin   0.5-0.7 m
Taiyuan   0.8 m   Dalian   0.8 m   Beijing   0.8-1.0 m
Shenyang   1.2 m   Changchun   1.6 m   Harbin   1.8-2.0 m

Manzhouli 2.8 m

ψ_{zs} ——the influence coefficient of soil type on freezing depth, which is shown in Table 3-4;

ψ_{zw} ——the influence coefficient of frost heaving of soil on freezing depth, which is shown in Table 3-5;

ψ_{ze} ——the influence coefficient of environment on freezing depth, which is shown in Table 3-6.

Coefficient of soil type on freezing depth　　Table 3-4

Types of soil	Influence coefficient ψ_{zs}	Types of soil	Influence coefficient ψ_{zs}
Cohesive soil	1.00	Medium, coarse and gravel sand	1.30
Fine sand, silt and silt	1.20	Gravel soil	1.40

Influence coefficient of frost heaving of soil on freezing depth　　Table 3-5

Frost heaving	Influence coefficient ψ_{zw}	Frost heaving	Influence coefficient ψ_{zw}
Non-frost heave	1.00	Strong frost heaving	0.85
Weak frost heaving	0.95	Extra strong frost heaving	0.80
Frost heaving	0.90		

Influence coefficient of environment on freezing depth　　Table 3-6

Surroundings	Influence coefficient ψ_{ze}	Surroundings	Influence coefficient ψ_{ze}
Village, town, wilderness	1.00	Urban area	0.90
Immediate vicinity of a city	0.95		

Note: For the environmental impact coefficient, when the urban population is 200,000-500,000, the value is taken according to the suburb. When the urban population is more than 500,000 and less than or equal to 1 million, the value is taken according to the urban area. When the urban population exceeds 1 million, the value is taken according to the urban area, and the suburban area within 5 km is taken according to the suburban area.

In seasonal frozen soil areas, the minimum buried depth of the foundation must be taken into consideration. If the foundation is too shallow, and there is a thick frost heaving soil layer underneath, the building may crack or not function properly due to freezing and thawing deformation of the soil. In cold areas, the ground freezing depth can be significant, requiring the foundation to be deeply buried. In reality, a thick frozen soil layer can exist beneath the basement, provided it does not produce excessive frost heaving force during freezing and does not result in excessive subsidence during thawing. The *Code for Design of Building Foundation* (*GB 50007—2011*) in China allows for a certain thickness of

Chapter 3 Design of Shallow Foundation on Natural Subgrade

frozen soil layer, based on systematic field tests and theoretical analyses, to ensure that the frost heave stress generated in the foundation during freezing does not exceed the additional stress caused by the external load at the corresponding position. Therefore, considering the frost heaving of foundation soil, the minimum buried depth of the foundation can be calculated using the following equation:

$$d_{min} = z_d - h_{max} \tag{3-3}$$

where

z_d ——freezing depth of foundation in seasonal frozen soil area;

h_{max} ——maximum thickness of allowable residual frozen soil layer under the foundation bottom surface, which can be found in Table 3-7.

Maximum allowable frozen soil thickness under the bottom of foundation h_{max} Table 3-7

Frost heave type	Foundation type	Heating situation	Average bearing pressure (kPa)					
			110	130	150	170	190	210
Weak frost heave soil	Square foundation	get heating	0.90	0.95	1.00	1.10	1.15	1.20
		no heating	0.70	0.80	0.95	1.00	1.05	1.10
	Strip foundation	get heating	>2.50	>2.50	>2.50	>2.50	>2.50	>2.50
		no heating	2.20	2.50	>2.50	>2.50	>2.50	>2.50
Frost heaving	Square foundation	get heating	0.65	0.70	0.75	0.80	0.85	
		no heating	0.55	0.60	0.65	0.70	0.75	
	Strip foundation	get heating	1.55	1.80	2.00	2.20	2.50	
		no heating	1.15	1.35	1.55	1.75	1.95	

Note: 1. This table only calculates the normal frost-heaving force of the foundation. If there is a tangential frost-heaving force on the side of the foundation, measures should be taken to prevent it.
2. The width of the foundation is less than 0.6 m, which is not applicable. The short-side dimension of the rectangular foundation is calculated according to the square foundation.
3. The data in the table are not applicable to silt, mucky soil and unconsolidated soil.
4. When calculating the average pressure of the basement, the standard combination value of permanent action should be multiplied by 0.9, which can be interpolated.

Example 3-1:

A city is located in the north with a population of 300,000, which gets heating. A three-story frame structure will be built. The subgrade soil is seasonal frost heaving silt soil layer with a standard frost depth of 2.4 m. The building adopts a squarespread foundation under the column, with side length of $b = 2.7$ m. Under the combined standard loads, the average pressure on the bottom of the foundation generated by the permanent load is 144.5 kPa. Try to determine the minimum buried depth of the foundation according to the frost heave requirements.

Solution:

According to Tables 3-4 to 3-6, $\psi_{zs} = 1.2$, $\psi_{zw} = 0.9$, $\psi_{ze} = 0.95$

$z_d = z_0 \, \psi_{zs} \, \psi_{zw} \, \psi_{ze} = 2.4 \times 1.2 \times 0.9 \times 0.95 = 2.4624$ m

The standard combination value of permanent load is 144.5 kPa. According to Table 3-7, average bearing pressure=0.9×144.5=130 kPa, $h_{max} = 0.70$ m

$$d_{min} = z_d - h_{max} = 2.4624 - 0.70 = 1.7624 \text{ m}$$

3.4 Subgrade bearing capacity verification

The bearing capacity of the subgrade refers to its capacity to bear the load transmitted from the foundation. When designing the foundation, it is important to ensure that the subgrade can safely support the load. The characteristic value of subgrade bearing capacity can be determined through load tests, in-situ tests, theoretical equations, and engineering practice experience.

3.4.1 Subgrade bearing capacity

1. Static load tests

In-situ static load tests are typically conducted to determine the characteristic value of subgrade bearing capacity. Its testing method has been described in Chapter 2. When determining the bearing capacity of subgrade by in-situ tests, the influence of the width and buried depth of foundation on bearing capacity is not considered. Only after the width and buried depth of foundation are corrected by the following equation can the characteristic value of bearing capacity of foundation be obtained for practical design.

$$f_a = f_{ak} + \eta_b \gamma (b-3) + \eta_d \gamma_m (d - 0.5) \tag{3-4}$$

where

f_a ——modified characteristic value of subgrade bearing capacity, kPa;

f_{ak} ——characteristic value of subgrade bearing capacity determined according to field load test or other in-situ tests and engineering experience, kPa;

γ ——natural gravity of soil below basement, and floating gravity below groundwater level, kN/m³;

γ_m ——weighted average weight of soil above the foundation bottom, and floating weight below the groundwater level, kN/m³;

b ——width of foundation, m. When the width is less than 3 m, it shall be counted as 3 m, and when it is greater than 6 m, it shall be counted as 6 m;

d ——buried depth of foundation, m, calculated generally from the outdoor ground elevation. In the fill levelling area, it can be calculated from the ground elevation of the fill, but when the fill is completed after the superstructure construction, it should be counted from the natural ground elevation. For a basement, if box foundation or raft foundation is used, it shall be calculated from the outdoor ground elevation. When using pad foundation or strip foundation, it should be calculated from the indoor ground elevation;

η_b, η_d —— the bearing capacity correction coefficient corresponding to the foundation width and buried depth, which is obtained according to the soil type under the bottom of foundation in Table 3-8.

Bearing capacity correction factor   Table 3-8

Types of soil			η_b	η_d
Silt and mucky soil			0	1.0
Artificial fill cohesive soil with e or I_L greater than or equal to 0.85			0	1.0
Red clay	moisture ratio, water content ratio $a_w > 0.8$		0	1.2
	moisture ratio, water content ratio $a_w \leqslant 0.8$		0.15	1.4
Large area Compacted fill	Silt with compaction coefficient greater than 0.95 and clay content $\rho_c \geqslant 10\%$		0	1.5
	Graded sand with maximum density greater than 2.1 t/m³		0	2.0
Silt	Silty clay with clay content $\rho_c \geqslant 10\%$		0.3	1.5
	Silty soil with grain content $\rho_c < 10\%$		0.5	2.0
Cohesive soil with e and I_L less than 0.85.			0.3	1.6
Silty sand and fine sand (excluding slightly dense state when it is very wet and saturated)			2.0	3.0
Medium sand, coarse sand, gravel sand and gravel soil			3.0	4.4

Note: 1. For strongly weathered and completely weathered rocks, values can be obtained by referring to the corresponding weathered soil types, and rocks in other states are not corrected.

2. Water content ratio $a_w = \dfrac{w}{w_L}$, natural water content w and liquid limit w_L.

3. Large-area compacted fill refers to the fill whose fill range is more than two times the width of the foundation.

2. Theoretical equation suggested by the code

If the eccentricity of the vertical force acting on the foundation is not more than 0.033 b (the width of the foundation), the foundation pressure is approximately evenly distributed. In this case, the characteristic value of the subgrade bearing capacity can be obtained by Equation 3-5.

$$f_a = M_b \gamma b + M_d \gamma_m d + M_c c_k \tag{3-5}$$

where

f_a —— characteristic value of subgrade bearing capacity which has taken into account the influence of foundation width and buried depth kPa;

c_k —— standard value of cohesion of soil, located under the foundation within the range of 1 time the width of the foundation, kPa;

3.4 Subgrade bearing capacity verification

M_b, M_d, M_c ——bearing capacity coefficients as shown in Table 3-9;

b ——the width of the foundation's bottom surface, which is taken as 6 m when it is greater than 6 m, and taken as 3 m when the width is less than 3 m for sand;

d, γ, γ_m ——same as Equation 3-4.

Bearing capacity coefficient M_b, M_d, M_c Table 3-9

Standard value of internal friction angle of soil (φ_k)	M_b	M_d	M_c	Standard value of internal friction angle of soil (φ_k)	M_b	M_d	M_c
0	0	1.00	3.14	22	0.61	3.44	6.04
2	0.03	1.12	3.32	24	0.80	3.87	6.45
4	0.06	1.25	3.51	26	1.10	4.37	6.90
6	0.10	1.39	3.71	28	1.40	4.93	7.40
8	0.14r	1.55	3.93	30	1.90	5.59	7.95
10	0.18	1.73	4.17	32	2.60	9.35	8.55
12	0.23	1.94	4.42	34	3.40	7.21	9.22
14	0.29	2.17	4.69	36	4.20	8.25	9.97
16	0.36	2.43	5.00	38	5.00	9.44	9.80
18	0.43	2.72	5.31	40	5.80	10.84	11.73
20	0.51	3.06	5.66				

Note: φ_k ——standard value of internal friction angle of soil within the range of 1 time the width of foundation under the bottom of foundation, (°).

3.4.2 Bearing capacity verification of bearing soil layer

The subgrade soil layer directly supporting the foundation is called the bearing soil layer, and it is required that the average bearing pressure acting on the bearing layer cannot exceed the bearing capacity of the soil layer, which is expressed as

$$p_k \leqslant f_a \tag{3-6}$$

where

p_k ——the average pressure value of the soil layer below the foundation, kPa;

f_a ——characteristic value of subgrade bearing capacity, kPa.

For all types of foundations, including raft foundations and box foundations, the foundation pressure is simplified to be distributed in a straight line when checking the bearing capacity of the foundation. This is determined using the principles of material mechanics. As Figure 3-15 shows, when the applied load is a central load, the calculation is as follows:

$$p_k = \frac{F_k + G_k}{A} \tag{3-7}$$

where

F_k ——the vertical force transmitted from the upper structure to the top surface of the foundation corresponding to the standard combination of the applied loads, kN;

G_k ——the weight of the foundation itself and the soil above the foundation, kN;

A ——bottom area of the foundation, m².

When the action is eccentric load, the distribution of the bearing pressure is presented in Figure 3-16. The maximum and minimum pressures can be calculated as follows:

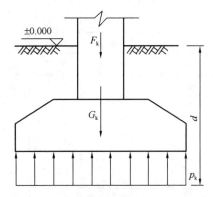
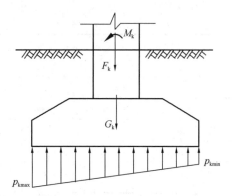

Figure 3-15 Distribution of bearing pressure under concentric vertical load

Figure 3-16 Distribution of bearing pressure under eccentric vertical load

$$p_{kmax} = \frac{F_k + G_k}{A} + \frac{M_k}{W} \tag{3-8}$$

$$p_{kmin} = \frac{F_k + G_k}{A} - \frac{M_k}{W} \tag{3-9}$$

$$M_k = (F_k + G_k)e \tag{3-10}$$

where

p_{kmax} ——maximum pressure at the bottom edge of foundation, kPa;

p_{kmin} ——minimum pressure at the bottom edge of foundation, kPa;

M_k ——moment acting on the bottom surface of the foundation corresponding to the standard combination, kN·m;

W ——resistance moment of foundation bottom surface, m³;

e ——eccentric moment of resultant force at the base, m.

When $e > b/6$ and $p_{kmin} < 0$, the bottom surface of one side of the foundation is separated from the subrade soil. In this case, the pressure of the foundation is distributed as shown in Figure 3-17. The p_{kmax} can be expressed as

$$p_{kmax} = \frac{2(F_k + G_k)}{3ac} \tag{3-11}$$

where

a ——side length of foundation bottom perpendicular to the direction of moment, m;

c ——the distance from the point of the resultant force to the maximum pressure edge of the foundation bottom surface, m.

For the bearing capacity verification under eccentric load, in addition to the requirements of Equation 3-7, it is also required that

$$p_{k\max} \leqslant 1.2 f_a \tag{3-12}$$

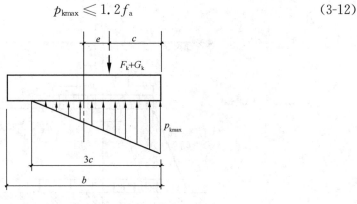

Figure 3-17 Distribution of bearing pressure when $e > b/6$

3.4.3 Bearing capacity verification of weak substratum

Below the bearing soil layer, if there is a soil layer whose strength and modulus are obviously lower than that of the bearing stratum, it is called weak substratum. If the soft substratum is not buried deep enough and the stress spreading to the substratum is greater than the bearing capacity of the substratum, the foundation may still fail, so it is necessary to check the bearing capacity of the soft substratum.

According to the elastic half-space theory, the stress on the top surface of the underlying layer is the largest at the central axis of the foundation, and it spreads around in a nonlinear distribution. If the properties of the upper and lower soils are considered, the stress distribution is more complicated and it is difficult to verify the bearing capacity. To simplify the calculation, it is generally assumed that the bearing pressure diffuses downward at a certain angle θ, as shown in Figure 3-18. The additional pressure p_z of strip foundation on the top surface of soft substratum is as follows:

$$p_z = \frac{b(p_k - p_{c0})}{b + 2z\tan\theta} \tag{3-13}$$

As for the rectangular foundation it is:

$$p_z = \frac{ab(p_k - p_{c0})}{(a + 2z\tan\theta)(b + 2z\tan\theta)} \tag{3-14}$$

where

b ——width of bottom surface of rectangular foundation or strip foundation, m;

a ——length of bottom surface of rectangular foundation, m;

p_k ——foundation bottom pressure, kPa;

p_{c0} ——self-weight pressure of soil at the bottom of foundation, kPa;

z ——the distance between the bottom surface of the foundation and the top surface of the soft substratum, m;

θ ——diffusion angle of bearing pressure, which can be found in Table 3-10.

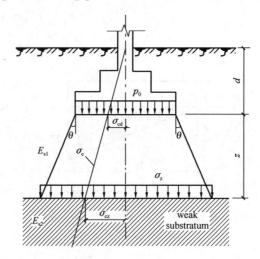

Figure 3-18 Bearing capacity verification of weak substratum

Diffusion angle of bearing pressure θ   **Table 3-10**

E_{s1}	z/b	
	0.25	0.50
3	6°	23°
5	10°	25°
10	20°	30°

Notes: 1. E_{s1} and E_{s2} are compressive modulus of upper soil and lower soil respectively.
    2. Generally, when $z < 0.25b$, take $\theta = 0°$, if necessary, it should be determined by an experiment. When $z > 0.50b$, θ is equivalent to that when $z = 0.50b$. When z/b is between 0.25 and 0.50, θ can be determined through interpolation.

The stress acting on the top surface of the weak substratum is not only the additional stress p_z, but also the self-weight stress at this depth p_{cz}, so the bearing capacity verification of the weak substratum should meet the following requirement:

$$p_{cz} + p_z \leqslant f_{d+z} \qquad (3-15)$$

where

f_{d+z} ——the characteristic value of subgrade bearing capacity after depth correction at the place where the top surface of weak substratum is embedded $d+z$, kPa.

If the bearing capacity of a weak substratum does not meet the requirements of Equation 3-15, the area of the foundation must be revised, and the bearing pressure shall be reduced until the requirements are met. In some cases, it may be necessary to modify the foundation scheme.

Example 3-2:

The base area of a foundation under column is 2 m×3 m, and the central load corresponding to the standard combination is $F_k = 800$ kN. The distribution of soil layers is shown in Figure 3-19. Try to verify the bearing capacity of the weak substratum of the foundation.

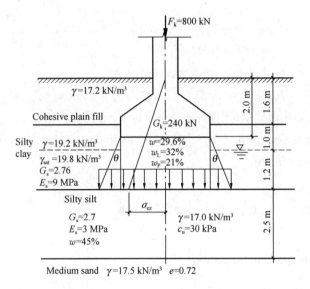

Figure 3-19 Distribution of soil layers

Solution:

(1) Calculate the bearing pressure

$$p_k = \frac{F_k + G_k}{A} = \frac{800 + 240}{2 \times 3} = 173.3 \text{kN/m}^3$$

(2) Calculate the bearing capacity of weak substratum

Use Equation 3-5

$$f_a = M_b \gamma b + M_d \gamma_m d + M_c c_k$$

$\varphi_k = 0° \Rightarrow M_b = 0, M_d = 1.0 \text{MPa}, M_c = 3.14 \text{MPa}$

The third layer of silty silt is a weak substratum. $z = 1.8$ m, $z/b = 0.9$, $E_{s1}/E_{s2} = 9/3 = 3 \Rightarrow \theta = 23°$

The buried depth of the stratum is $d = 1.6 + 2.2 = 3.8$ m

The weighted average weight of the soil above the weak substratum is

$$\gamma_m = (1.6 \times 17.2 + 1.0 \times 19.2 + 1.2 \times 10)/3.8 = 15.5 \text{kN/m}^3$$

then $f_a = 3.14 \times 30 + 1.0 \times 15.5 \times 3.8 = 153$ kPa

(3) Calculate the pressure acting on weak substratum

Self-weight pressure at the top surface of the weak substratum:

$$p_{cz} = 17.2 \times 1.6 + 19.2 \times 1.0 + (19.8 - 9.8) \times 1.2 = 58.72 \text{kN/m}^2$$

Additional stress on the bottom of the foundation:

Chapter 3 Design of Shallow Foundation on Natural Subgrade

$$p_z = \frac{ab(p_k - p_{co})}{(a + 2z\tan\theta)(b + 2z\tan\theta)}$$

$$p_{co} = 17.2 \times 1.6 + 19.2 \times 0.4 = 35.2 \text{kN/m}^2$$

$$p_k - p_{co} = 173.3 - 35.2 = 138.1 \text{kN/m}^2$$

then

$$p_z = \frac{138.1 \times 3 \times 2}{(3 + 2 \times 1.8 \times \tan 23°)(2 + 2 \times 1.8 \tan 23°)}$$

$$= 51.82 \text{kN/m}^2$$

Total pressure on underlying weak stratum:

$$p_{cz} + p_z = 58.72 + 51.82 = 110.54 \text{kN/m}^2$$

(4) Verification of the underlying stratum bearing capacity

It can be found that $f_a > p_{cz} + p_z$, therefore the underlying weak substratum meets the bearing capacity requirement.

3.5 Evaluation of subgrade settlement

3.5.1 Allowable subgrade deformation

In the ultimate state design of foundations, the evaluation of subgrade deformation is of utmost importance. In principle, all types of buildings must comply with the requirements of the following equation:

$$s \leqslant [s] \tag{3-16}$$

However, for many buildings with simple geological conditions, few floors, and small loads, there is enough engineering experience showing that the above-mentioned bearing capacity requirements are met, which also meets the foundation deformation requirements. Therefore, the *Code for the Design of Building Foundation* (GB 50007—2011) specifies the scope of foundation deformation evaluation as follows: Grade A and Grade B buildings should be designed according to foundation deformation requirements. For Grade C buildings within the range listed in Table 3-1, deformation evaluation are not required. However, deformation evaluation should be carried out if they fall within the range listed in Table 3-1 and meet one of the following conditions.

(1) Buildings with complex shapes and the characteristic value of subgrade bearing capacity is less than 130 kPa.

(2) If the foundation is subject to ground surcharge in its vicinity, or if there is a significant load difference between adjacent foundations, the risk of excessive uneven settlement may arise.

(3) When there is eccentric load on the building on the soft foundation.

(4) When the adjacent buildings are close and may tilt.

(5) When there is filled soil with large thickness or uneven thickness in the foundation, its self-weight consolidation is not completed.

Foundation settlement caused by deformation can be classified into four categories: settlement, differential settlement, tilt and local tilt, as depicted in Figure 3-20. The type of settlement varies depending on the type of building. Typically, masonry load-bearing structures are governed by the local tilt values, while frame structures and single-layer bent structures are governed by the differential settlement of adjacent column foundations. Multi-storey and high-rise buildings are primarily controlled by the tilt value, and the average settlement should be evaluated if necessary.

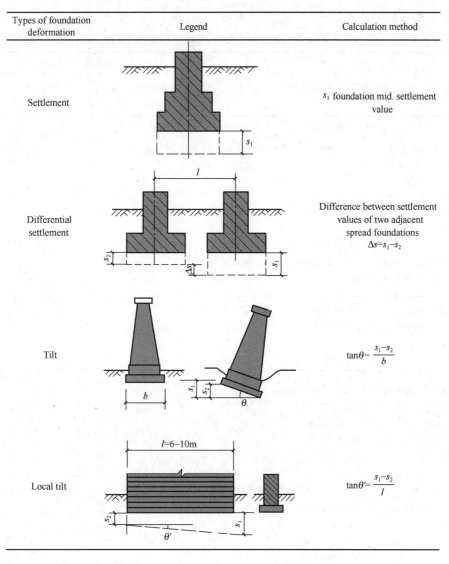

Figure 3-20 Foundation settlement classification

The impact of foundation deformation varies depending on the structural type and intended use of the building. Different types of buildings have different sensitivities to foundation deformation, and the damage caused by deformation can differ in its impact on use function. The *Code for the Design of Building Foundation* (GB 50007—2011) includes a large number of existing buildings that have been observed for settlement and investigated for their use. Based on geological conditions, Table 3-11 provides allowable values for foundation deformation of various buildings, which can be used for engineering analysis and application. For buildings not included in the table, allowable values for foundation deformation can be determined based on the adaptability of the superstructure to foundation deformation and the requirements of use.

Allowable subgrade deformation of buildings [s] Table 3-11

Deformation characteristics	Foundation soil category	
	Medium and low compressibility soil	High compressibility soil
Local inclination of masonry load-bearing structure foundation	0.002	0.003
Settlement difference between adjacent column foundations of industrial and civil buildings Δs (1) Frame structure (2) Side columns filled with masonry walls (3) Structure that does not generate additional stress when the foundation is unevenly settled	0.002 l 0.0007 l 0.005 l	0.003 l 0.001 l 0.005 l
Settlement of column foundation of single-layer bent structure (column spacing is 6 m) s/mm	(120)	200
Inclination of rail surface of bridge crane (According to not adjusting the track) vertical crosswise		0.004 0.003
Overall inclination of multi-storey and high-rise buildings $H_g \leqslant 24$ $24 < H_g \leqslant 60$ $60 < H_g \leqslant 100$ $H_g > 100$		0.004 0.003 0.0025 0.002
Average settlement of foundation of simple high-rise building /mm		200
Inclination of high-rise structure foundation $H_g \leqslant 20$ $20 < H_g \leqslant 50$ $50 < H_g \leqslant 100$ $100 < H_g \leqslant 150$ $150 < H_g \leqslant 200$ $200 < H_g \leqslant 250$		0.008 0.006 0.005 0.004 0.003 0.002

3.5 Evaluation of subgrade settlement

continued

Deformation characteristics	Foundation soil category	
	Medium and low compressibility soil	High compressibility soil
Settlement of high-rise structure foundation s/mm $H_g \leqslant 100$ $100 < H_g \leqslant 200$ $200 < H_g \leqslant 250$		400 300 200

Note: 1. Brackets are only applicable to moderately compressible soil.
      2. The values in this table are the allowable values of the actual final deformation of the building foundation.

3.5.2 Calculation of subgrade settlement

The subgrade settlement is a complicated problem influencedby many the factors, which is described in detail in the soil mechanics textbook. The *Code for Design of Building Foundation* (GB 50007—2011) summarizes a lot of engineering experience, and it is suggested that the layered summation method be used for calculation. The method is shown in Figure 3-21 and its expression is as follow:

$$s = \psi_s s' = \psi_s \sum_{i=1}^{n} \frac{p_0}{E_{si}} (\bar{a}_i z_i - \bar{a}_{i-1} z_{i-1}) \tag{3-17}$$

where

s ——final deformation of foundation, mm;

s' ——calculated settlement of foundation, mm;

ψ_s ——empirical coefficient of settlement calculation, which is determined according to the regional experience and can be obtained in Table 3 12;

p_0 ——additional bearing pressure under quasi-permanent combination, kPa;

z_i, z_{i-1} ——depth of the bottom surface and the top surface of the ith layer of soil respectively;

$\bar{a}_i$, $\bar{a}_{i-1}$ ——the average additional stress coefficients, which can be found from Table 3-13 according to the length-width ratio a/b and depth ratio z/b of the foundation.

When using Table 3-12 to obtain the empirical coefficient of settlementcal culation ψ_s, $\overline{E}_s$ value is the equivalent compressive modulus within the depth range of deformation calculation. It can be calculated as follow:

$$\overline{E}_s = \frac{\Sigma A_i}{\Sigma \frac{A_i}{E_{si}}} \tag{3-18}$$

where

A_i ——the area of additional stress distribution of the ith soil layer.

Chapter 3 Design of Shallow Foundation on Natural Subgrade

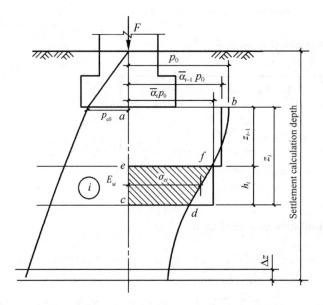

Figure 3-21 Subgrade settlement calculation by layered summation method

Empirical Coefficient of Settlement Calculation ψ_s   Table 3-12

Additional bearing pressure at base of foundation	$\overline{E}_s$/MPa				
	2.5	4.0	7.0	15.0	20.0
$p_0 \geq f_{ak}$	1.4	1.3	1.0	0.4	0.2
$p_0 \leq 0.75 f_{ak}$	1.1	1.0	0.7	0.4	0.2

In the layered summation method, the settlement depth is generally calculated according to Equation 3-19.

$$\Delta s' \leq 0.025 s' \qquad (3\text{-}19)$$

Average additional stress coefficient along cenerline of a rectangular under uniformly distributed load in rectangular area $\overline{\alpha}$ Table 3-13

| z/b | a/b | | | | | | | | | | | |
	1.0	1.2	1.4	1.6	1.8	2.0	2.4	2.8	3.2	3.6	4.0	5.0	>10.0 Strip foundation
0.0	1.000	1.000	1.000	1.000	1.000	1.000	1.000	1.000	1.000	1.000	1.000	1.000	1.000
0.2	0.987	0.990	0.991	0.992	0.992	0.992	0.993	0.993	0.993	0.993	0.993	0.993	0.993
0.4	0.936	0.947	0.953	0.956	0.958	0.960	0.961	0.962	0.962	0.963	0.963	0.963	0.963
0.6	0.858	0.878	0.890	0.898	0.903	0.096	0.910	0.912	0.913	0.914	0.914	0.915	0.915
0.8	0.775	0.801	0.810	0.831	0.839	0.844	0.851	0.855	0.857	0.858	0.859	0.860	0.860
1.0	0.598	0.738	0.749	0.764	0.775	0.783	0.792	0.798	0.801	0.803	0.804	0.806	0.807
1.2	0.631	0.663	0.686	0.703	0.715	0.725	0.737	0.744	0.749	0.752	0.754	0.756	0.758
1.4	0.573	0.605	0.629	0.648	0.661	0.672	0.687	0.696	0.701	0.705	0.708	0.711	0.714

3.6 Evaluation of subgrade stability

continued

z/b	a/b												>10.0 Strip foundation
	1.0	1.2	1.4	1.6	1.8	2.0	2.4	2.8	3.2	3.6	4.0	5.0	
1.6	0.524	0.556	0.580	0.599	0.613	0.625	0.641	0.651	0.658	0.663	0.666	0.670	0.675
1.8	0.482	0.513	0.537	0.556	0.571	0.583	0.600	0.611	0.619	0.624	0.629	0.633	0.638
2.0	0.446	0.475	0.499	0.518	0.533	0.545	0.563	0.575	0.584	0.590	0.594	0.600	0.606
2.2	0.414	0.443	0.466	0.484	0.499	0.511	0.530	0.543	0.552	0.558	0.563	0.570	0.577
2.4	0.387	0.414	0.436	0.454	0.4.69	0.481	0.500	0.513	0.523	0.530	0.535	0.543	0.551
2.6	0.362	0.389	0.410	0.428	0.442	0.455	0.473	0.487	0.496	0.504	0.509	0.518	0.528
2.8	0.341	0.366	0.387	0.404	0.418	0.430	0.449	0.463	0.472	0.480	0.486	0.495	0.506
3.0	0.322	0.346	0.366	0.383	0.397	0.409	0.427	0.441	0.451	0.459	0.465	0.474	0.487
3.2	0.305	0.328	0.348	0.364	0.377	0.389	0.407	0.420	0.431	0.439	0.445	0.455	0.468
3.4	0.289	0.312	0.331	0.346	0.359	0.371	0.388	0.402	0.412	0.420	0.427	0.437	0.452
3.6	0.276	0.297	0.315	0.330	0.343	0.353	0.372	0.385	0.395	0.403	0.410	0.421	0.436
3.8	0.263	0.284	0.301	0.316	0.328	0.339	0.356	0.369	0.379	0.388	0.394	0.405	0.422
4.0	0.251	0.271	0.288	0.302	0.314	0.325	0.342	0.355	0.365	0.373	0.379	0.391	0.408
4.2	0.241	0.260	0.276	0.290	0.300	0.312	0.328	0.341	0.352	0.359	0.366	0.377	0.396
4.4	0.231	0.250	0.265	0.278	0.290	0.300	0.316	0.329	0.339	0.347	0.353	0.365	0.384
4.6	0.222	0.240	0.255	0.268	0.279	0.289	0.305	0.317	0.327	0.335	0.341	0.353	0.373
4.8	0.214	0.231	0.245	0.258	0.269	0.279	0.294	0.300	0.316	0.324	0.330	0.342	0.362
5.0	0.206	0.223	0.237	0.249	0.260	0.269	0.284	0.296	0.306	0.313	0.320	0.332	0.352

Note: a, b —— long side and short side of rectangle;

z —— depth from the plane of load action.

where

$\Delta s'$ —— calculated settlement within the thickness of Δz (Figure 3-21), which can be obtained in Table 3-14, mm.

Value of Δz Table 3-14

b/m	$\leqslant 2$	$2 < b \leqslant 4$	$4 < b \leqslant 8$	$b > 8$
Δz/m	0.3	0.6	0.8	1.0

3.6 Evaluation of subgrade stability

It is rare for vertical load to cause foundation instability, so it is not necessary to check the stability of subgrade for general buildings, which can meet the bearing capacity. For buildings that often bear horizontal loads, such as hydraulic structures, retaining structures, high-rise buildings and towering structures, the subgrade stability may become the main problem in design. It is necessary to check the subgrade stability. Under the combined action of horizontal and vertical loads, there are two forms of subgrade instability and failure: one is surface sliding along the bottom surface of the foundation, as shown in Figure 3-22(a); The other is deep overall sliding failure, as shown in Figure 3-22(b).

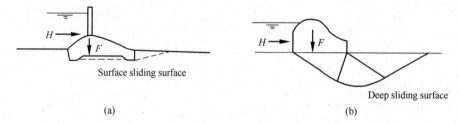

Figure 3-22 Failure form of foundation under inclined load

At present, the method of single safety factor is still used to check the stability of foundation. When it is determined that it belongs to surface sliding, the stability safety factor can be calculated by Equation 3-20:

$$F_s = \frac{f \cdot F}{H} \quad (3-20)$$

where

F_s ——safety factor of surface sliding related to the building grade, it is generally 1.2-1.4;

F ——the sum of vertical forces acting on the base, including the vertical components of the self-weight of the structure and other loads, kN;

H ——the sum of horizontal load components acting on the foundation, kN;

f ——the friction coefficient between foundation and foundation soil, which can be selected with reference to Table 3-15.

Friction coefficient between foundation and foundation soil　　Table 3-15

Types of soil		friction factor f
Cohesive soil	Plastic	0.25-0.30
	Hard plastic	0.30-0.35
	Hard	0.35-0.45
Silt		0.30-0.40
Medium sand, coarse sand and gravel sand		0.40-0.50
Gravel soil		0.40-0.60
Soft rock		0.40-0.60
Hard rock with rough surface		0.65-0.75

When the foundation instability belongs to deep sliding, the circular sliding method can be used for checking calculation. The stability safety factor refers to the ratio of the anti-sliding moment and sliding moment generated by the forces acting on the most dangerous sliding surface to the sliding centre, and its value shall meet the requirements of the following equation:

$$F_s = \frac{M_R}{M_s} \geq 1.2 \qquad (3-21)$$

where

M_R ——anti-sliding moment, kN·m;

M_s ——sliding moment, kN·m.

For the circular sliding method, please refer to the relevant contents of the soil mechanics textbook.

For the foundation stability of the building on the top of the soil slope, it is first necessary to check whether the soil slope itself is stable. If the soil quality of the slope is good and uniform, and the groundwater level is low, there will be no groundwater escaping from the foot of the slope, then the safe slope angle can be found in Table 3-16. At the same time, buildings should be prevented from being too close to the free surface of the slope, so as to prevent the foundation load from destabilizing the slope. Therefore, the horizontal distance a from the outer edge line of the foundation bottom surface to the top of the slope is required to meet the conditions of Equation 3-22 and Equation 3-23, and shall not be less than 2.5 m.

Allowable values of soil slope gradient   Table 3-16

Types of soil	Density or state	Allowable value of slope (aspect ratio)	
		The slope height is within 5 m	The slope height is 5-10 m
Gravel soil	Dense, medium and slightly dense	1:0.50-1:0.35	1:0.75-1:0.50
		1:0.75-1:0.50	1:1.00-1:0.75
		1:1.00-1:0.75	1:1.25-1:1.00
Cohesive soil	Hard and hard plastic	1:1.00-1:0.75	1:1.25-1:1.00
		1:1.25-1:1.00	1:1.50-1:1.25

Note: 1. Gravel soil in the table is filled with cohesive soil in hard or hard plastic state.
2. For gravel soil with sandy soil or sandy soil as filler, the allowable values of slope gradient are determined according to the natural angle of repose.

For strip foundations $\qquad a \geq 3.5b - \dfrac{d}{\tan\beta} \qquad (3-22)$

For rectangular foundations $\qquad a \geq 2.5b - \dfrac{d}{\tan\beta} \qquad (3-23)$

where

b —— the side length of the bottom surface of the foundation perpendicular to the edge line of the top of the slope, m;

d ——buried depth of foundation, m;

β ——slope angle, as shown in Figure 3-23.

When the height of the soil slope is too large and the slope angle is too steep, which is not within the applicable range of Table 3-16, or the requirements of Equation 3-22 or Equation 3-23 cannot be met due to the limitation of building layout, the overall stability of the slope and the foundation of the building on it should be checked by circular sliding method or other similar slope stability analysis methods.

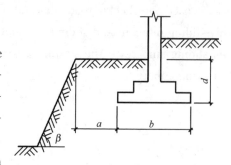

Figure 3-23   Horizontal distance from the top of the slope to the outer edge of the foundation's bottom

3.7 Design of unreinforced spread footings

A pad or strip foundation bears the load from the column or wall above, and its working condition is similar to that of an inverted cantilever beam with two sides extending outwards. This type of foundation is prone to bending failure at the edge of the column, wall, or step with a sudden change in section height. To prevent bending damage, foundations made of materials with poor tensile properties, such as bricks, masonry, plain concrete, lime soil, and concrete, must have a certain height to ensure that the tensile stress generated by bending does not exceed the tensile strength of the materials. Typically, the ratio of the overhanging length of the foundation to the height of the foundation should not exceed the specified allowable ratio. Table 3-17 provides the allowable values of various materials. The height H_0 of the foundation shall meet the requirements of the following equation (Figure 3-24):

$$H_0 = \frac{b - b_0}{2\tan\alpha} \qquad (3-24)$$

Allowable width-height ratio of unreinforced spread foundations   Table 3-17

Foundation material	Quality requirements	Allowable value of step width-height ratio		
		$p_k \leqslant 100\text{kPa}$	$100\text{kPa} < p_k \leqslant 200\text{kPa}$	$200\text{kPa} < p_k \leqslant 300\text{kPa}$
Concrete foundation	C15 concrete	1:1.00	1:1.00	1:1.25
Rubble concrete foundation	C15 concrete	1:1.00	1:1.25	1:1.50
Brick foundation	Brick is not less than MU10. Mortar is not less than M5	1:1.50	1:1.50	1:1.50
Rubble foundation	Mortar is not less than M5	1:1.25	1:1.50	—

continued

Foundation material	Quality requirements	Allowable value of step width-height ratio		
		$p_k \leqslant 100$kPa	100kPa $< p_k \leqslant 200$kPa	200kPa $< p_k \leqslant 300$kPa
Lime soil foundation	Minimum dry density of lime soil with volume ratio of 3 : 7 or 2 : 8: Silt 1.55 t/m³ Silty clay 1.50 t/m³ Clay 1.45 t/m³	1 : 1.25	1 : 1.50	—
Concrete foundation	The volume ratio is 1 : 2 : 4-1 : 3 : 6 (lime: sand: aggregate), and each layer is about 220mm virtual paving and rammed to 150mm	1 : 1.50	1 : 2.00	—

Note: 1. p_k refers to the average pressure at the bottom of the foundation when the standard combination is applied, kPa.
2. The width of each step of stepped rubble foundation should not be greater than 200 mm.
3. When the foundation is composed of different materials, the contact part shall be checked for compressive strength.
4. The concrete foundation with an average pressure value of more than 300 kPa at the bottom of the foundation should be checked for shear resistance.

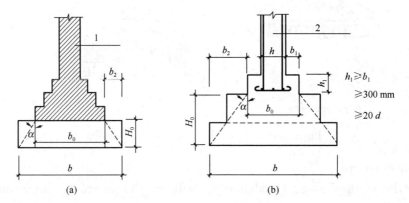

Figure 3-24 Schematic view of unreinforced spread foundations

3.8 Design of spread footings

3.8.1 Pad foundations

Spread footing is a type of reinforced concrete components that is subjected to both bending and shearing forces. Under the action of loads, the following failure forms may occur (Figure 3-25).

(1) Punching failure

The research of reinforced concrete shows that under the combined action of bending and shear loads, the main failure form of structures begins with oblique cracks appearing in the bending-shear area first, with an increase in load, the cracks expand upward, and the normal stress and shear stress of the uncracked part increase rapidly. The combination of normal stress and shear stress causes the tensile stress, which is greater than the tensile strength of concrete, the oblique cracks are pull apart and oblique tensile failure occurs, which is also called punching failure on the spread foundation (Figure 3-25a). In general, the punching failure controls the height of the spread foundations.

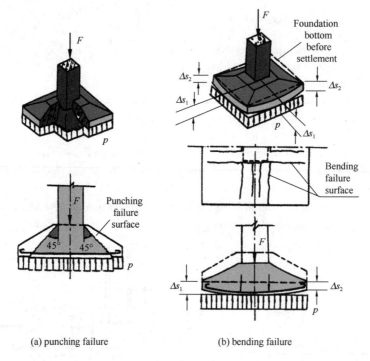

(a) punching failure   (b) bending failure

Figure 3-25   Failure form of spread foundation

(2) Shear failure

When the width of a pad foundation is small and the punching failure cone may fall outside the foundation, shear failure may occur along the vertical plane at the junction of the column and the foundation or at the region the step changes.

(3) Bending failure

The stratum reaction produces bending moment at the foundation surface. When the bending moment becomes excessive, foundation bending failure occurs. This failure occurs along the wall, column or step edge, and the crack is parallel to the wall or column edge (Figure 3-25b). In order to prevent this kind of damage, it is required that the bending moment M generated by the base reaction on each vertical section of the foundation is less than or equal to the bending strength M_u of this section, and the reinforcement of the foun-

3.8 Design of spread footings

dation is determined according to this condition in design.

(4) When the concrete strength grade of the foundation is less than that of the column, the top surface of the foundation may be damaged by local compression.

Therefore, when designing spread footing, the following items should be checked.

1. Punching failure verification of pad foundation

The pad foundation often bears eccentric load, and the base reaction force is unevenly distributed. Figure 3-26 shows schematic diagram of punching failure calculation for pad foundations. It can be expressed as follows:

$$F_l \leqslant 0.7\beta_{hp} f_t a_m h_0 \quad (3\text{-}25)$$
$$a_m = (a_t + a_b)/2 \quad (3\text{-}26)$$
$$F_l = p_j A_l \quad (3\text{-}27)$$

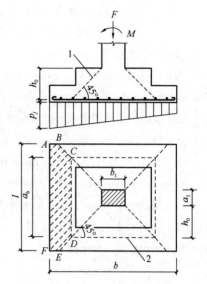

Figure 3-26 Schematic diagram of punching failure calculation

where

F_l ——outside the range of base punching cone, the punching load caused by net pressure p_j on the failure cone;

A_l ——part of the basement area used in punching calculation, (shaded area ABCDEF in Figure 3-26), m²;

p_j ——corresponding to the basic combination of actions, the net reaction force per unit area of foundation soil, and the maximum net reaction force per unit area (kPa) of foundation soil at the edge of eccentric compression foundation $p_{j\max}$ can be taken;

β_{hp} ——punching height influence coefficient, when $h \leqslant 800$ mm, take $\beta_{hp} = 1.0$, when $h \geqslant 2,000$ mm, take $\beta_{hp} = 0.9$, when 800mm$<h<$2000mm, calculate β_{hp} according to linear interpolation method;

f_t ——the design tensile strength of concrete;

a_m ——calculated length of the most unfavourable side of the punch-breaking cone, m;

a_b ——the length of the upper side of the inclined section on the most unfavorable side of the punching failure cone, and the width of the column shall be taken when calculating the punching bearing capacity of the junction between the column and the foundation, m;

a_t ——the length of the bottom of the inclined section at the most unfavourable side of the punching failure cone which is within the area of the foundation bottom. When the bottom of the punching failure cone falls within the foundation bottom, calculating the punching bearing capacity at the junction of the column and the foundation, requires taking the column width plus two times the effective height of the foundation, m;

h_0 ——the effective height of the foundation, m.

The punching of a stepped pad foundation can be checked using the similar method. However, it should be noted that the above calculation only considers the strength of one inclined section of the cone, while ignoring the beneficial influence of the strength of the other side. Despite this, the calculation method is considered safe.

2. Shear failure calculation of pad foundation

When the width of the pad foundation bottom surface is less than or equal to the column width plus twice the effective height of the foundation, the shear bearing capacity of the section at the junction of the column and the foundation shall be checked according to the following equations:

$$V_s \leqslant 0.7\beta_{hs} f_t A_0 \tag{3-28}$$

$$\beta_{hs} = (800/h_0)^{1/4} \tag{3-29}$$

where

V_s ——design value of shear force at the junction of column and foundation, and the shaded area $ABCD$ in Figure 3-27 is multiplied by the average net reaction force of the base p_j, kN;

β_{hs} ——shear height influence coefficient. When $h_0 <800$ mm, take $h_0 =800$ mm; when $h_0 > 2,000$ mm, take $h_0 =2,000$ mm;

A_0 ——effective vertical area along the BD, m², as shown in Figure 3-27. $A_0 = l \times h_0$.

3. Bending failure calculation of pad foundation

Pad foundation is subjected to reaction from the base, resulting in two-way bending. In the analysis, the foundation can be divided into four regions according to the connecting line from the corners of the column to the four vertices of the foundation base. The bending moment is maximum at sections I-I and II-II along the column edge (Figure 3-28). The bending moments caused by the pressure acting on the bottom surface to the I-I and II-II section can be obtained as follows:

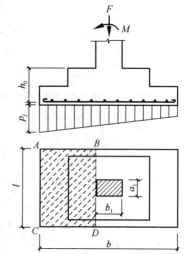

Figure 3-27 Schematic diagram of shear failure calculation

$$M_\mathrm{I} = \frac{1}{12}a_1^2 \left[(2l+a')\left(p_{max}+p-\frac{2G}{A}\right) + (p_{max}-p)l \right] \tag{3-30}$$

$$M_{\mathrm{II}} = \frac{1}{48}(l-a')^2(2b+b')\left(p_{\max}+p_{\min}-\frac{2G}{A}\right)$$

(3-31)

After the bending moment of each section of the foundation is obtained, the area of reinforcing steel required by the foundation can be calculated according to Equation 3-32.

$$A_s = \frac{M}{0.9f_y h_0} \qquad (3\text{-}32)$$

where

f_y ——tensile strength design value of steel bar.

3.8.2 Strip footings under wall

The width of the strip-shaped spread foundation under the wall can be determined in the same way as that of the pad foundation. According to previous engineering experience, the foundation

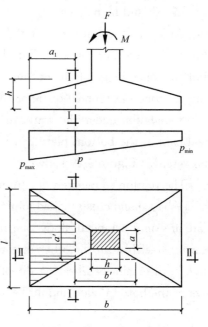

Figure 3-28 Schematic diagram of bending failure calculation

height can be taken as 1/8 of the foundation width, and then determined by shear checking calculation. Generally, the unit length, i.e. 1m, is taken for checking the shear and bending resistance. When checking, calculate the shear force and bending moment of dangerous section (such as the foot of the wall or the step change) according to the net pressure distribution of the basement. The foundation height should be calculated according to the verification of shear bearing capacity, which is shown as follow:

$$V_s = 0.7\beta_{\mathrm{hs}} f_t h_0 \qquad (3\text{-}33)$$

3.8.3 Structural requirement of spread footings

Cast-in-place under-column spread footings are generally made into cone and step shapes. The height of the edge of the conical foundation should be not less than 200 mm, and the slope in both directions should be less than or equal to 1 : 3. The height of each step of the stepped foundation should be 300-500 mm. The thickness of the cushion should be not less than 70 mm, and the strength grade of the cushion concrete should be not less than C15. The minimum reinforcement ratio of the load-bearing steel bars should be not less than 0.15%, the minimum diameter of the bottom plate load-bearing steel bars should be not less than 10 mm, and the spacing should be less than or equal to 200 mm and not less than 100 mm. The diameter of the longitudinally distributed steel bars in the reinforced concrete strip foundation under the wall should be not less than 8 mm, and the spacing should be less than or equal to 300 mm; The area of steel bars distributed per line-

Chapter 3 Design of Shallow Foundation on Natural Subgrade

ar meter should be not less than 15% of the area of stressed steel bars. When there is a cushion layer, the thickness of the steel bar protective layer should be not less than 40 mm, and when there is no cushion layer, it should be not less than 70 mm. The strength grade of concrete should be not less than C20. When the side length of the independent reinforced concrete foundation under the column and the width of the reinforced concrete strip foundation under the wall are not less than 2.5 m, the length of the load-bearing steel bars in the bottom plate can be 0.9 times the side length or width, and should be staggered, as Figure 3-29 shown.

The reinforced concrete strip foundation bottom plate is located at the intersection of the T-shaped and cross shaped joints. The transverse load-bearing steel bars of the bottom plate are only arranged along one main load-bearing direction, and the transverse load-bearing steel bars in the other direction can be arranged at 1/4 of the width of the bottom plate in the main load-bearing direction. At the corner, the transverse load-bearing steel bars of the bottom plate should be arranged in both directions.

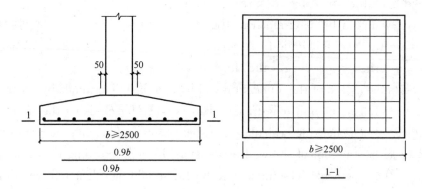

Figure 3-29 Arrangement of load-bearing steel bars

Example 3-3:

A 5-story building is located at the top of a slope, with an angle between the slope surface and the horizontal plane $\beta = 45°$. The upper structure adopts a reinforced concrete frame structure, with a pad foundation under the column, The centreline of the foundation bottom surface coincides with the centreline of the column section. The cross-sectional size of the column is taken as 500 mm × 500 mm, the shape of the foundation bottom is square with its width $b = 2.5$ m. The foundation profile and soil layer distribution are shown in the Figure 3-30. The weighted unit weight of the foundation and above soil is 20 kN/m³, and there is no groundwater in the site. The partial coefficient of the weight of the foundation and its upper soil is taken as 1.35. Corresponding to the basic load, the vertical force $F = 1600$ kN and a unidirectional moment M_x are applied on the top surface of the foundation, and the minimum reaction design value $p_{min} = 230$ kPa. Try to determine the maximum bending moment (kN · m) of the foundation at the column edge.

3.9 Methods to reduce damages due to uneven settlement of buildings

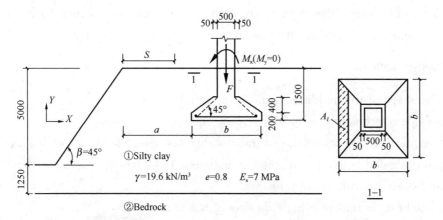

Figure 3-30 The foundation and the distribution of soil layers (mm)

Solution:

(1) Transfer the weight of the foundation and above soil to the basis value

$G = 1.35 G_k = 1.35 \times 20 \times 2.5^2 \times 1.5 = 253.125 \text{ kN}$

(2) Calculate the bearing pressure under base combination

$$p = (F+G)/A = (1,600+253.125)/(2.5 \times 2.5) = 296.5 \text{ kPa}$$

$$p_{\max} = 2p - p_{\min} = 2 \times 296.5 - 230 = 363 \text{ kPa}$$

$$a_1 = (2.5 - 0.5)/2 = 1.0 \text{ m}$$

$p = p_{\min} + (p_{\max} - p_{\min})(b - a_1)/b = 230 + (363 - 230) \times (2.5 - 1)/2.5 = 309.8 \text{ kPa}$

(3) Calculate the bending moment of the foundation at the edge of column

$$M_I = \frac{1}{12} a_1^2 \left[(2l + a') \left(p_{\max} + p - \frac{2G}{A} \right) + (p_{\max} - p)l \right]$$

$$= \frac{1}{12} \times 1.0^2 \times \left[(2 \times 2.5 + 0.5)\left(363 + 309.8 - \frac{2 \times 253.125}{2.5 \times 2.5}\right) \right.$$

$$\left. + (363 - 309.8) \times 2.5 \right]$$

$$= 282.3 \text{ kN} \cdot \text{m}$$

3.9 Methods to reduce damages due to uneven settlement of buildings

Uneven settlement of buildings can occur due to uneven subgrade or large differences in the loads of superstructures. If the uneven settlement exceeds the allowable limit, the buildings can crack, sustain damage, and even pose serious harm.

The mitigation of harm caused by uneven settlement is a crucial aspect of building construction design. Since the upper structure, foundation, and subgrade interact andfunction collectively, the design work should incorporate comprehensive technical measures to minimize the harm caused by uneven settlement.

3.9.1 Design methods

1. The shape of the building should be as simple as possible

The shape of a building includes its plane and elevation shapes, as well as its dimensions. A simple building with high overall rigidity can resist deformation better. Therefore, when designing buildings on soft foundations, it is advisable to use a simple shape, such as an "I" shape with uniform height.

The height or weight of a building should not vary significantly, as this can lead to excessive uneven settlement caused by different loads in the parts with abrupt height changes. Surveys have shown that masonry bearing structure houses with a height difference of more than one floor on soft soil foundations are prone to cracking (as shown in Figure 3-31). Therefore, in areas with weak foundations, building designs should avoid height differences of more than one floor.

2. Control the length-height ratio of buildings

The ratio of the length of a building on the plane to the height from the bottom of the foundation is known as the length-height ratio. Masonry buildings with a high length-height ratio have lower overall rigidity, and

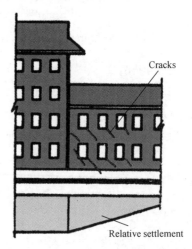

Figure 3-31 Cracks due to the high elevation difference

their longitudinal walls are more likely to crack due to excessive deflection, as shown in Figure 3-32. For buildings with three or more floors, if the estimated maximum settlement exceeds 120 mm, the length-height ratio should not exceed 2.5. For buildings with a simple plan, well-connected interior and exterior walls, and small spacing between transverse walls, the length-height ratio is usually not more than 3.0. If the ratio exceeds 3.0, settlement joints should be installed.

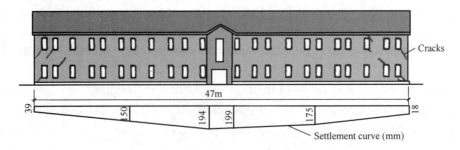

Figure 3-32 Cracks due to excessive length-height ratio

The reasonable arrangement of vertical and horizontal walls is an essential measure to improve the overall stiffness of masonry buildings. In case of a weak foundation, it is advisable to keep the internal and external vertical walls from turning or minimize their turning, maintain appropriate spacing between the internal and horizontal walls, and ensure a firm connection with the vertical walls. If required, the rigidity and strength of the foundation should be reinforced.

3. Design settlement joints

When the shape of a building is complex or the ratio of length to height is too large, settlement joints can be installed from the foundation to divide the building into independent settlement units. Typically, each unit should have a simple shape, a small length-height ratio, the same structural type, and uniform foundation, so that each settlement unit has high overall stiffness and relatively uniform settlement, which will prevent further cracking.

Generally, settlement joints should be set in the following parts of the building.

(1) The turning of a building.

(2) Places where there is a big difference in height or load of buildings.

(3) Masonry structure with unsatisfactory length ratio and proper parts of reinforced concrete frame structure.

(4) The compressibility of foundation soil has obvious changes.

(5) Different types of structures or foundations.

(6) The junction of building houses by stages.

(7) Expansion joints are to be set, which can also be used as settlement joints.

4. Control the net distance between adjacent building foundations

Due to the diffusion and superposition of additional stress in the foundation, on the weak foundation, the adjacent foundations that are too close to each other will influence the settlement of the foundation, and the resulting additional uneven settlement may cause the cracking or mutual inclination of the building. This adjacent influence is mainly manifested as follows.

(1) Two adjacent buildings built in the same period will affect each other, especially when there is a big difference between the weight of the two buildings, the light (low) one is greatly affected by the heavy (high) one.

(2) The original buildings are affected by the new heavy or high-rise buildings nearby.

The required clear distance between adjacent building foundations can be selected according to Table 3-18 It can be seen from the table that the main indexes determining the clear distance between foundations are the stiffness (measured by the ratio of length to height) of the affected building and the estimated average settlement of the affected building. The latter comprehensively reflects the compressibility of the foundation, the scale and weight of the affected building and other factors.

Net distance between adjacent building foundations Table 3-18

Estimated average settlement of new buildings (s/mm)	The length-height ratio of the affected building (m)	
	$2.0 \leqslant L/H_f < 3.0$	$3.0 \leqslant L/H_f < 5.0$
70-150	2-3	3-6
160-250	3-6	6-9
260-400	6-9	9-12
>400	9-12	$\geqslant 12$

Note: 1. In the table, the unit length separated by houses or settlement joints is L, and H_f is the height of houses from the elevation of foundation bottom, m.

2. When the length-height ratio of the affected building is $1.5 < L/H_f < 2.0$, the clear distance between the foundations can be appropriately reduced.

The distance between the outer walls of adjacent high-rise structures (or structures with strict requirements on tilt) can be calculated and determined according to the allowable value of tilt.

5. Adjust the design elevation of buildings

If the settlement of a building is excessive, it can cause changes to the building's original elevation. This, in turn, can result in damage to pipelines, rainwater leakage, obstruction of equipment operations, and other issues that can affect the building's normal use. To address this, adjustments can be made using appropriate measures such as the follows.

(1) According to the estimated settlement, appropriately raise the elevation of indoor floor or underground facilities.

(2) When there is a connection between various parts of the building (or equipment), the elevation of those with larger settlement can be appropriately increased.

(3) There should be enough space between the building structure and installed equipment.

(4) When pipelines pass through buildings, holes of sufficient size should be reserved, or flexible pipe joints should be used.

3.9.2 Structural methods

1. Reduce the self-weight of a building

The self-weight of the building (including the weight of foundation and overlying soil) accounts for a large proportion of the foundation pressure, which is estimated to be about 1/2 for industrial buildings and more than 3/5 for civil buildings. Therefore, reducing the self-weight of buildings can effectively reduce the foundation settlement. Measures taken to achieve this include the follows.

(1) Reduce the weight of the wall, such as hollow blocks, perforated bricks or other

light wall materials.

(2) Choose light structure, such as prestressed concrete structure, light steel structure and various light space structures.

(3) Reduce the weight of foundation and backfill soil on it. You can choose the foundation form with less covering soil and light self-weight, such as compensatory foundation and shallow buried reinforced expanded foundation. If the indoor floor is high, the raised floor can be used to reduce the thickness of indoor backfill soil.

2. Set ring beam

The purpose of using ring beams is to increase the bending resistance of masonry structures and improve the overall bending stiffness of the building. This is an effective method to prevent cracks in brick walls and prevent them from expanding. In the event of uneven settlement, the wall may deflect positively or negatively, and the lower or upper ring beams will help to counteract this deflection. Ring beams are typically installed both above and below the building.

Ring section, reinforcement and plane layout, etc., can be combined with the requirements of the code for seismic design of buildings. Multi-storey houses should be set near the foundation surface and at the top of the top doors and windows, and other floors can be set by layers. When the local foundation is weak, or the building shape is complex and the load difference is large, it can be set layer by layer.

Ring beams must be integrated with masonry, and each ring beam should try to penetrate all external walls, load-bearing internal longitudinal walls and main internal transverse walls to form a closed system on the plane. If the wall is severely weakened due to the large opening and the foundation is weak, it may be considered to properly reinforce the weakened part or strengthen it with reinforced concrete frame.

3. Set up foundation beam (ground beam)

Reinforced concrete frame structure is very sensitive to uneven settlement. For frame structure with single column foundation, setting foundation beam between foundations is one of the effective measures to increase structural rigidity and reduce uneven settlement. The bottom of the foundation beam is generally placed on the foundation surface (or slightly higher), the section height can be $1/14$-$1/8$ of the column spacing, and the reinforcement is evenly distributed from top to bottom, and the reinforcement ratio on each side is 0.4%-1.0%.

4. Reduce that additional pressure of the adjust substrate

(1) Set the basement (or semi-basement). Compensatory foundation design method is adopted to offset part or even all of the building weight by the excavated soil weight, so as to reduce the additional pressure and settlement of the foundation. Basement (or semi-basement) can also be set only in the part of the building where the load is particularly large. By this method, the settlement of each part of the building tends to be uniform.

(2) Adjust the base size. After determining the size of the bottom of the foundation according to the bearing capacity of the foundation, the settlement can be reduced by applying the settlement theory and necessary calculation and combining with the design experience, adjusting the size of the foundation and increasing the bottom area of the foundation.

5. Adopt a structural type that is insensitive to differential settlement

Masonry bearing structure and reinforced concrete frame structure are very sensitive to uneven settlement, while hinged structures such as bent frame and three-hinged arch (frame) have great adaptability to uneven settlement. The relative displacement of supports will not cause great additional stress, which can avoid the harm of uneven settlement. However, the hinged structure is usually only suitable for single-storey industrial plants, warehouses and some public buildings. Flexible base plates are often used for the foundation slabs of oil tanks and pools, so as to better adapt to uneven settlement.

3.9.3 Construction methods

It is very important to adopt a reasonable construction sequence and method when carrying out engineering construction on soft foundation, which is one of the effective measures to reduce or adjust uneven settlement.

1. Arrange the construction sequence reasonably

When the proposed adjacent buildings are different in weight and height, generally, the construction should be carried out according to the procedure of "first heavy and then light, first high and then low". If necessary, the light adjacent buildings should be built after the completion of the high-heavy buildings. If the heavy main building is connected with the light subsidiary part, it should also be handled according to the above principles.

2. Pay attention to construction methods

Around the completed buildings, it is not advisable to pile up a large number of building materials or earthworks, so as to avoid additional settlement of buildings caused by ground load.

If there are any buildings with pile foundations in the proposed dense buildings, the pile construction should be carried out first, and attention should be paid to adopting a reasonable pile sinking sequence.

When lowering the groundwater level and excavating deep foundation pits, we should pay close attention to the possible adverse effects on adjacent buildings, and if necessary, we can take measures such as setting water cut-off curtain and controlling the deformation of foundation pits.

When excavating the foundation trench on the highly sensitive silt and muddy soft soil foundation, attention should be paid to protecting the bearing layer from disturbance. Usually, the original soil layer with a thickness of about 200 mm is reserved at the bottom of the pit, and the concrete cushion is temporarily excavated manually. If the soft soil at the bottom of the pit is disturbed, the disturbed part can be excavated and backfilled with sand, gravel (brick), etc. . During the construction in rainy period, the soil at the bottom of the pit should be prevented from being soaked by rainwater. Also pay attention to control the loading rate.

Questions

3-1 How many types of foundation are there? What are their characteristics and the scopes of application?

3-2 What is buried depth of the foundation? Please describe its influencing factors.

3-3 What factors determine the freezing depth of foundation and how to determine the minimum burial depth of the foundation?

3-4 What method can be used to determine the characteristic bearing capacity of subgrade?

3-5 How to correct the influence of width and depth of the foundation on the bearing capacity of the subgrade?

3-6 What requirements need to be met for checking the bearing capacity of the foundation bearing layer?

3-7 How to verify the bearing capacity of the weak substratum of the foundation?

3-8 What is the spread foundation? What are the advantages compared to rigid foundations?

3-9 What is the punching failure of the foundation? How to perform punching failure verification?

3-10 What are the types of foundation deformation in buildings? Give an example to illustrate the deformation control for a certain type of buildings.

3-11 How to reduce uneven settlement from the layout of buildings?

3-12 What structural measures can be taken to reduce uneven settlement of buildings?

3-13 What are the construction methods to reduce the damage caused by uneven settlement of buildings?

3-14 A high-rise residential building adopts a raft foundation with the size of 12 m × 50 m. The average bearing pressure corresponding to the standard combination of load at the bottom of the foundation is 325 kPa. The distribution of the soil layer is shown in the Figure 3-33. Try to determine the corrected characteristic bearing capacity of soil at the bottom of the foundation.

3-15 A strip foundation is under the reinforced concrete wall, the foundation profile and soil layer distribution are shown in Figure 3-34. The average bearing pressure under the standard combination of load at the bottom of the foundation per unit length is 250 kN/m. The weighted unit weight of soil and foundation is 20 kN/m^3, and the diffusion angle to the weak substratum is $\theta = 12°$. Please determine the with of the strip foundation according to the requirement of bearing capacity.

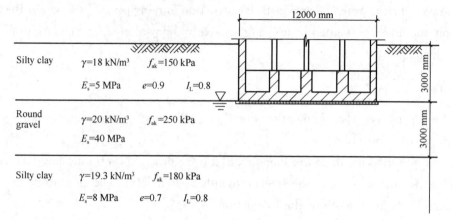

Figure 3-33　Distribution of the soil layers

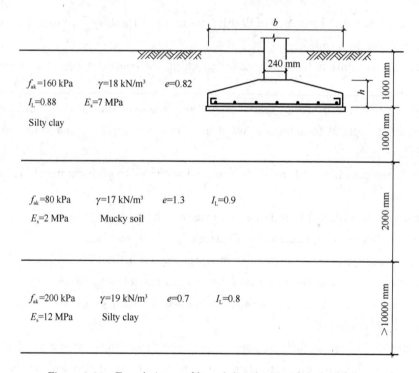

Figure 3-34　Foundation profile and distribution of the soil layers

Chapter 4 Continuous Foundations

4.1 Introduction

Currently various types of high-rise buildings are required to build in complex geological conditions with the development of China's economy. The design and construction of foundations for these high-rise buildings are the key technical and economic conditions for ensuring the stability, safety and normal use of these buildings. Continuous foundations, such as strip foundations, raft foundations and box foundations, are usually chosen to use as the main foundation types of this kind of heavy buildings due to the following advantages of these foundations: ① They have a large bottom area and their high bearing capacity is easy to meet the requirements of subgrade bearing capacity; ② The continuity of these foundations can greatly enhance the overall stiffness of buildings, which is beneficial for reducing uneven settlement and improving the seismic performance of buildings; ③ Box foundations and raft foundations with basements can improve the bearing capacity of the foundation and compensate for some or all of the weight of the building with the excavated soil.

In the design calculation and analysis of rigid foundations and spread foundations, the superstructure, subgrade and foundation are simply divided into three independent components according to the static balance condition. This simplification is usually safe and applicable because the internal force and deformation errors produced are not big. However, for the strip foundations, raft foundations and box foundations, because of their large size, deep depth and great load, the upper part is integrated with the structure, and the lower part is closely combined with the subgrade soil. The superstructure, subgrade and foundation work together. If the superstructure, subgrade and foundation are simply separated, only the static balance conditions are satisfied without considering the influence of the deformation coordination conditions among them, it will often cause great errors and even get incorrect results. Therefore, compared with rigid foundation and spread foundation, when designing continuous foundations, superstructure, subgrade and foundation should not only meet the conditions of static balance, but also meet the conditions of deformation compatibility.

In order to solve interactions among the superstructure, subgrade and foundation, the followings should be considered in the design of continuous foundations.

(1) Establish a subgrade model that can better reflect the deformation characteristics of subgrade soil and the method to determine model parameters. In this way, the stiffness of the subgrade soil can be quantitatively calculated in the interaction analysis.

(2) Establish the interaction theory among the superstructure, subgrade and foundation. According to the stiffnesses of superstructure, subgrade and foundation, the deformation can be calculated to satisfy deformation compatibility.

4.2 Interaction of subgrade, foundation and superstructure

4.2.1 Interaction concept of subgrade, foundation and superstructure

The superstructure is connected with the foundation through walls and columns, and the bottom surface of the foundation is in direct contact with the subgrade, which constitutes a complete system. Their contact points not only transfer loads, but also constrain and interact with each other. The interaction effect among the subgrade, foundation and superstructure mainly depends on their stiffness.

1. Interaction between the subgrade and foundation

The stiffness of the subgrade is the ability of the subgrade to resist deformation, which is manifested in the softness or compressibility of soil. If the subgrade soil is incompressible, the foundation will not flex, and the superstructure will not generate additional internal force due to the uneven settlement of the foundation. In this case, the interaction of the joint action is very weak, and the superstructure, subgrade and foundation can be calculated separately, which is close to the situation for buildings with rock foundation and dense gravel and sand foundation, as shown in Figure 4-1(b). Generally, the subgrade soil hasa certain compressibility. Under the condition that the rigidity of the superstructure and foundation is constant, the weaker the subgrade soil is, the greater the relative deflection and internal force of the foundation will be, and the corresponding secondary stress will be caused to the superstructure, as shown in Figure 4-1(a).

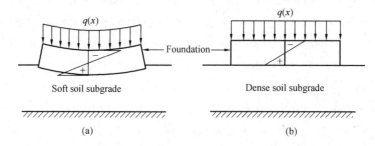

Figure 4-1 Influence of different compressible subgrade on foundation deflection and internal force

As shown in Figure 4-2, the uneven distribution of the compressed soil layer in the foundation is obvious, and two different forms of uneven distribution will produce two completely different results on the deflection and internal force of the foundation and superstructure.

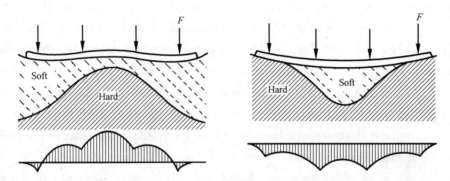

Figure 4-2 Influence of uneven subgrade on foundation deflection and internal force

The foundation transfers the load of the superstructure to the subgrade. In this process, through its own stiffness, it adjusts the load of the superstructure upward and restricts the deformation of the subgrade downward, so that the superstructure, subgrade and foundation form a whole system to satisfy stress and deformation compatibility. The foundation plays a key role in the working of the system. For the convenience of analysis, regardless of the role of the superstructure, assuming that the foundation is completely flexible, then the load transfer is not constrained by the foundation and has no diffusion effect, the distributed load acting on the foundation will be directly transmitted to the subgrade, resulting in the subgrade reaction force with the same size and equal load distribution. When the load $q(x,y)$ is evenly distributed, the reaction force $p(x,y)$ is evenly distributed, as shown in Figure 4-3(a). However, the uniformly distributed load on the subgrade will cause the concave deformation of the surface as shown in the figure. Obviously, in order to make the foundation settlement uniform, the load and the subgrade reaction force must be distributed in a parabolic shape with small middle and large sides, as shown in Figure 4-3(b).

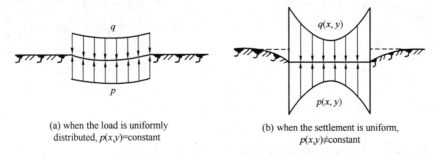

(a) when the load is uniformly distributed, $p(x,y)$=constant

(b) when the settlement is uniform, $p(x,y) \neq$ constant

Figure 4-3 Subgrade reaction of flexible foundation

Rigid foundation should restrain and adjust the load transfer and foundation deformation. Assuming that the foundation is absolutely rigid, there is a uniform load acting on it. In order to adapt to the inflexible characteristics of the absolutely rigid foundation, the subgrade reaction force will concentrate on both sides, forcing the foundation surface to deform evenly to adapt to the settlement of the foundation. When the subgrade soil is regarded as a complete elastic body, the reaction force distribution of the foundation will be parabolic as shown in Figure 4-4(a). The actual subgrade soil has only limited strength, and the stress at the edge of the foundation is too great, so the soil will yield and undergo plastic deformation, and part of the stress will shift to the middle. The distribution of reaction force is as shown in Figure 4-4(b), that is, the saddle-shaped distribution. As far as the ability to bear shear stress is concerned, the soil in the middle part of the foundation is higher than the soil at the edge, so when the load continues to increase, the damage range of the soil at the edge below the foundation continues to expand, and the reaction force further shifts from the edge to the middle. It becomes the bell-shaped distribution as shown in Figure 4-4(c). If the subgrade soil is cohesionless and has no cohesive strength, and the buried depth of the foundation is very shallow, and the dead weight pressure on the outside of the edge is very small, the soil at this place has almost no shear strength, so it can't bear any load, so the distribution of reaction force may become the distribution of inverted parabola as shown in Figure 4-4(d).

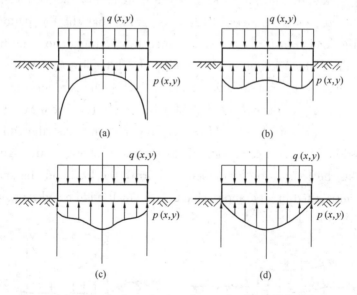

Figure 4-4 Distribution of subgrade reaction of rigid foundation

If the foundation is not an absolute rigid body, but a finite rigid body, the foundation will flex to a certain extent under the combined action of the load from the superstructure and the foundation reaction, and the subgrade soil will deform accordingly under the action of the foundation reaction. The real distribution curve of reaction force depends on the relative stiffness of subgrade and foundation. The greater the stiffness of the foundation and the smaller the stiffness of the subgrade, the higher the concentration of the subgrade reaction to the edge.

2. The interaction between superstructure and foundation

Regardless of the influence of the subgrade, it is assumed that the subgrade is deformed and the reaction force on the bottom of the foundation is evenly distributed, as shown in Figure 4-5(a). If the superstructure is an absolutely rigid body (such as cast-in-place shear wall structure with great rigidity) and the foundation is a strip or raft foundation with small rigidity, when the foundation is deformed, because the superstructure does not bend, the columns can only sink evenly, and the constrained foundation cannot occur. In this case, the foundation is like an inverted continuous beam supported on bearings with the column end as a fixed hinge, with the subgrade reaction as the load, and only local bending occurs between bearings.

As shown in Figure 4-5(b), if the superstructure is a flexible structure (such as a frame structure with less overall rigidity) and the foundation is also a strip or raft foundation with less rigidity, the superstructure has no or only a small constraint on the deformation of the foundation. Therefore, the foundation not only bends locally due to the reaction of the foundation between spans, but also bends as a whole with the deformation of the structure, and the superposition of the two will produce greater deformation and internal force.

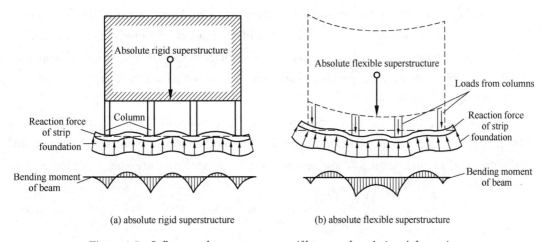

(a) absolute rigid superstructure   (b) absolute flexible superstructure

Figure 4-5 Influence of superstructure stiffness on foundation deformation

If the stiffness of the superstructure is between the above two extreme cases, it is obvious that with the increase of the stiffness of the superstructure, the deflection and internal force of the foundation will decrease. At the same time, the superstructure will produce secondary stress due to the displacement of the column end. Further analysis shows that if the foundation also has a certain stiffness, the deformation and internal force of the superstructure and foundation must be affected by their stiffness, which can be analyzed through the distribution of internal force at the joints, which belongs to structural mechanics and will not be elaborated in this chapter.

3. Interaction of superstructure, subgrade and foundation

To analyze the deflection and internal force of the superstructure considering the interaction among the superstructure, subgrade and foundation, the key problem is to solve the distribution of the subgrade reaction force. However, it is very complicated to solve the actual subgrade reaction distribution because the real subgrade reaction distribution is restricted by the requirement of subgrade-foundation deformation compatibility. The deflection of the foundation depends on the load (including the subgrade reaction) acting on it and its own stiffness. The deformation of subgrade surface depends on all ground loads (i.e. subgrade reaction) and soil properties. Even if the subgrade soil is regarded as an ideal elastic material, it is not simple to calculate the reaction force distribution by using the deformation compatibility conditions of foundation displacement at each point, not to mention that the soil is not an ideal elastic material. Up to now, the interaction among the superstructure, subgrade and foundation can be solved in principle, but there is no perfect method to give satisfactory answers to all kinds of foundation conditions in practice. The most important difficulty is to choose the correct subgrade model.

4.2.2 Subgrade models

In the interaction analysis among superstructure, subgrade and foundation, or in the analysis of beam and slab on subgrade, the relationship between force and displacement at the contact interface between subgrade and foundation should be used. This relationship can be expressed by a continuous or discrete function, which is the so-called subgrade model. Usually there are three kinds of subgrade models which will be detailed as follows.

1. Winkle subgrade model

Winkler subgrade model was put forward by E. Wencker in 1867. The model assumes that the deformation at any point on the surface of subgrade soil s_i is directly proportional to the pressure at that point p_i, and has nothing to do with the pressure at other points, that is

$$p_i = k s_i \tag{4-1}$$

where

k ——resistance coefficient of subgrade, also called subgrade coefficient, kN/m^3.

Winkle subgrade model regards the subgrade as a series of soil columns with no friction on the side on a rigid subgrade. It can be simulated by a series of independent springs, as shown in Figure 4-6(a). The deformation of the subgrade is proportional to the pressure only in the load area, and the deformation outside the area is zero. The distribution pattern of subgrade reaction force is similar to the vertical displacement pattern of subgrade surface. Obviously, when the stiffness of the subgrade is large and it does not deflect after being stressed, the subgrade reaction force is distributed in a straight line according to the assumption of Winkler subgrade, as shown in Figure 4-6(c). When subjected to central load, it is uniformly distributed.

In fact, the subgrade is a very broad continuous medium, and the deformation of any point on the surface depends not only on the load directly acting on that point, but also on the whole ground load. Therefore, the actual subgrade that strictly conforms to the Winkler subgrade model does not exist. However, for the soft soil subgrade with low shear strength, or the subgrade compression layer is thin, it can be considered to conform to the Winkler subgrade model. For other cases, the application of Winkler subgrade model will produce larger errors. Winkle subgrade model is simple in expression and convenient in application, so it currently has been widely used in the design of strip, raft and box foundations.

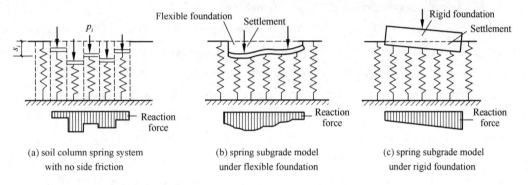

Figure 4-6　Schematic diagram of Winkle subgrade model

2. Elastic half-space subgrade model

The elastic half-space subgrade model regards the subgrade as a homogeneous linear deformation half-space, and uses the elastic mechanics theory to solve the stress or displacement in the subgrade. The settlement of any point on the subgrade is related to the whole subgrade reaction force and the adjacent load distribution.

According to Boussinesq's solution, when a vertical concentrated force acts on the surface of an elastic half space, the settlements of the subgrade surface at the distance r from the action point of vertical concentrated force on the surface is:

$$s = \frac{1-\nu^2}{\pi E} \cdot \frac{P}{r} \tag{4-2}$$

where

ν ——Poisson's ratio of soil;

E ——deformation modulus of soil.

For a continuous load distributed over a limited area, the deformation of each point on the subgrade surface can be obtained by integrating the basic Equation 4-2. For example, the load evenly distributed in the rectangular area, the deformation at the center of the rectangle is obtained through integration of Equation 4-2, which is given in the following:

$$s = \frac{2(1-\nu^2)}{\pi E}\left[l\ln\frac{b+\sqrt{l^2+b^2}}{l} + b\ln\frac{l+\sqrt{l^2+b^2}}{b}\right]p \qquad (4\text{-}3)$$

where

l, b——the length and width of the rectangular.

The elastic half-space subgrade model has the advantage of diffusing stress and deformation, and it can reflect the influence of adjacent loads. However, its diffusing capacity often exceeds the actual subgrade situation. And the calculated settlement and the settlement range of the ground surface are often larger than the measured results. At the same time, the model cannot take into account the important factors such as the layering and heterogeneity of the subgrade and the nonlinearity of the stress-strain relationship of the soil.

3. Finite compression layer subgrade model

When the subgrade soil layer distribution is complex, it is difficult to simulate using the above Winkler subgrade model or elastic half-space subgrade model. In this case, it is more appropriate to adopt the finite compression layer subgrade model.

In the finite compression layer subgrade model, the subgrade is regarded as a limited depth compression layer under the lateral limit condition, and the relationship between the deformation of the compression layer and the acting load of the foundation is established based on the layerwise summation method. Its advantage is that the subgrade can be layered, and the subgrade soil is assumed to be compressed under the condition of complete lateral restraint. So it is easy to obtain the compression modulus of the subgrade soil E_s by using the subgrade parameters through the field or indoor tests. The thickness of compression layer is still determined according to the layerwise summation method.

4.3 Strip footings under columns

4.3.1 Structure requirement and construction

Strip foundation under columns is a common foundation type on soft foundation, which can be divided into strip foundation beams extending in one direction (Figure 3-3) and cross foundation beams extending in two orthogonal directions (Figure 3-4).

(1) The strip foundation under the column is usually a reinforced concrete beam, which is composed of a rectangular rib beam in the middle and wing plates extending to both sides. It forms an inverted T-beam structure with large longitudinal bending stiffness and large base area. The typical structure is shown in Figure 4-7(d).

(2) In order to increase the bottom area of the beam foundation, improve its bearing conditions, and at the same time adjust the center of gravity of the base to coincide or be close to the center of load, both ends of the beam foundation should extend out of the side

column for a certain length (Figure 4-7a) l_0, which is generally 0.25 time of the side span l_1, that is $l_0 \leqslant 0.25l_1$. In this way, the reaction force distribution of the base can be more uniform and reasonable, thus reducing the bending effect.

(3) In order to improve the longitudinal bending stiffness of the strip foundation beam n and ensure a large enough base area, the cross section of the foundation beam is usually inverted T-shaped (Figure 4-7d). The beam height h is determined according to the bending calculation, which is generally 1/8-1/4 of the column spacing. The width of the wing plate protruding from the bottom is determined by the bearing capacity of the foundation, and the thickness of the wing plate h' is determined by the transverse bending calculation of the beam section, which should generally not be less than 200mm. When the thickness of the wing plate is greater than 250 mm, it should be made into a thick plate, and the top slope of the thick plate should be taken $i \leqslant 1/3$.

(4) The longitudinal section of the strip foundation beam is generally equal. In order to ensure the reliable connection with the column end, the beam width should be slightly larger than the column side length in this direction. If the short side of the column bottom section is perpendicular to the beam axis direction, the width of the rib beam is 50 mm wider than that of the column side. If the long side of the column bottom section is perpendicular to the beam axis direction, and the side length is 600 mm or greater than or equal to the width of the rib beam, it is necessary to widen the rib beam locally, and the distance from the edge of the column to the edge of the foundation shall not be less than 50 mm (Figure 4-7e).

(5) The stress of the foundation beam under the column is complex, which is not only subjected to longitudinal overall bending, but also to local bending between columns. After the superposition of the two factors, it is difficult to determine the actual bending moment direction in the column support and the span between columns completely according to the calculation. Therefore, the upper and lower sides of the beam are usually equipped with longitudinal reinforced bars (Figure 4-7b and c), and the reinforcement ratio of each side is not less than 0.2%. In addition to meeting the calculation requirements, the longitudinal reinforced bars at the top and bottom should be fully penetrated according to the calculated reinforcement number, and the length reinforced bars at the bottom should not be less than 1/3 of the total area of reinforced bars at the bottom. The reinforced bars of the bearing under the column in the foundation beam should be arranged at the lower part of the bearing, and the reinforced bars between columns should be arranged at the middle and upper parts of the span. The lap position of the lower longitudinal reinforcement of the beam should be in the middle of the span, while the lap position of the upper longitudinal reinforcement of the beam should be at the bearing, and both should meet the requirements of lap length.

(6) When the height of the beam is greater than 700 mm, structural waist bars with a

diameter greater than 10 mm should be added at intervals of 300-400 mm along the height on both sides of the beam, and the stirrups of ribbed beams should be closed with a diameter of not less than 8 mm (Figure 4-7d). The number of bent tendons and stirrups shall be configured according to bending moment and shear force diagram. When the width of the beam is $b \leqslant 350$ mm, use double limb hoops, four limb hoops when $b > 350$ mm, and six limb hoops when $b > 800$ mm. The spacing of stirrups is the same as that of ordinary beams.

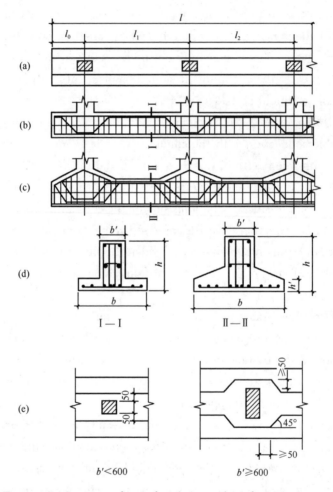

Figure 4-7 Structure of strip foundation under column (unit: mm)

(7) The concrete strength grade of the reinforced concrete foundation beam under the column is generally not lower than C20, and the bottom surface of the foundation beam in the soft soil area should be provided with a sand cushion with a thickness of not less than 100 mm. If plain concrete cushion is used, the general strength grade is C7.5, and the thickness is not less than 75 mm. When the concrete strength grade of the foundation beam is less than that of the column, the local compressive strength of the top surface of the foundation beam under the column should be checked.

4.3.2 Internal force calculation

It should be noted that the internal force calculation of the strip foundation under the column and the strip foundation under the wall is totally different. For the strip foundation under the wall, the load of the wall is longitudinally distributed continuously, which can usually be regarded as longitudinally distributed load. Therefore, the strip foundation under the wall can be analyzed by taking a single-width cross section as a plane problem. However, the column load borne by the strip foundation under the column can be considered as concentrated load, which is evenly or unevenly distributed on several nodes of the foundation beam. Under the joint action of column load and foundation reaction, the foundation beam will produce longitudinal deflection, so it is necessary to analyze the internal force of the whole beam.

As mentioned above, the key to the internal force analysis of strip foundation under column is how to determine the subgrade reaction distribution. Considering the interaction among superstructure-subgrade-foundation, the calculation is very complicated. In order to learn the basic theory of foundation design and understand several practical calculation methods, the commonly used inverted beam method without considering the interaction will be presented as follows.

Inverted beam method is a foundation beam method without considering the interaction of superstructure-foundation-foundation. It is suitable for the following situations: 1) the stiffness of the superstructure and foundation are relatively large; 2) the height of the foundation beam is not less than 1/6 of the column spacing; 3) the superstructure load distribution is relatively uniform, that is, there is little difference between the column spacing and the column load; 4) the soil layer distribution and soil quality of the foundation are relatively uniform. These conditions make the deflection of the foundation beam very small, and the reaction force on the bottom of the foundation basically conforms to the linear distribution. It can be considered that there is no mutual constraint between the superstructure-subgrade-foundation. The relationship among them only needs to meet the static equilibrium condition, without considering the deformation compatibility. At this time, due to the large stiffness of the superstructure, there will be no obvious displacement difference at the column foot. The foundation beam is like an inverted multi-span continuous beam with the upper side fixedly hinged to the column end and the lower side subjected to the linear distributed foundation reaction (Figure 4-8b). The structural mechanics method can be applied, that is, the moment distribution method or the empirical moment coefficient method can be directly applied to solve the internal force of the foundation beam, so it is called the inverted beam method.

The calculation steps of inverted beam method are as follows.

(1) According to the foundation size determined by foundation calculation, the basic

combination acting in the limit state of bearing capacity is used to calculate the internal force of foundation.

(2) Calculate the net reaction distribution of the foundation (Figure 4-8a). In the calculation of the subgrade reaction, the self-weight of the foundation is not counted. It is considered that the self-weight of the foundation will not cause internal forces in the foundation beam. In the following description, the self-weight of the foundation beam is ignored. So the subgrade reaction p also represents the net reaction p_j. The subgrade net reaction can be calculated as follows:

$$p_{\substack{jmax \\ jmin}} = \frac{\sum F}{bL} \pm \frac{\sum M}{W} \tag{4-4}$$

where

$p_{\substack{jmax \\ jmin}}$ ——maximum and minimum net reaction force of base, kPa;

$\sum F$ ——sum of vertical loads design values, kN;

$\sum M$ ——the sum of the design values of the external bending moment on the centroid of the basement, kN·m;

W ——resistance moment of basement area, $W = \frac{1}{6}bL^2$, m³;

b, L ——width and length of the bottom surface of foundation beam, m.

(3) Determine the calculation diagram. With the column end as the fixed hinge support and the net reaction of the basement as the load, draw the calculation diagram of multi-span continuous beam, as shown in Figure 4-8(a). If the actual situation is considered, the interaction between the superstructure and the foundation will cause the arch effect, that is, the end foundation reaction will increase during the deformation of the foundation. So the foundation reaction should be increased by 15%-20% at both ends of the strip foundation, as shown in Figure 4-8(b).

(4) Calculate the bending moment distribution (Figure 4-8c), shear force distribution (Figure 4-8d) of continuous beam and the bearing reaction force R_i by using bending moment distribution method or other solutions.

(5) Adjust and eliminate the unbalanced force of the bearing. Obviously, the first calculated reaction force of the bearing is usually not equal to the column load, which can't meet the static equilibrium condition at the bearing. The reason is that in this calculation, it is assumed that the column foot is fixed compared with the bearing and the subgrade reaction force is distributed in a straight line, which can't be met at the same time. For the unbalanced force, it needs to be eliminated through successive adjustment. The adjustment method is as follows.

① Obtain the unbalanced force ΔP_i according to the column load at the bearing F_i and the bearing reaction force R_i.

$$\Delta P_i = F_i - R_i \quad (4\text{-}5)$$

② Convert the bearing unbalanced force into distributed load Δq, which is evenly distributed in the 1/3 span range of two adjacent spans of the bearing.

For the side span bearing,

$$\Delta q_i = \frac{\Delta P_i}{l_0 + \dfrac{l_i}{3}} \quad (4\text{-}6)$$

For the mid-span bearing,

$$\Delta q_i = \frac{\Delta P_i}{\dfrac{l_{i-1}}{3} + \dfrac{l_i}{3}} \quad (4\text{-}7)$$

where

Δq_i ——uniformly distributed load converted from unbalanced force, kN/m^2;

l_0 ——extension length of foundation beam under side column, m;

l_{i-1}, l_i ——left and right span length of the bearing, m.

Apply the distributed load on the continuous beam, as shown in Figure 4-8(e).

③ Calculate the bending moment ΔM, shear force ΔV and bearing reaction of continuous beam ΔR_i under the action Δq. Superimpose ΔR_i on the original bearing reaction R_i, thus obtaining a new bearing reaction $R'_i = R_i + \Delta R_i$. If R'_i is close to the column load F_i and the difference is

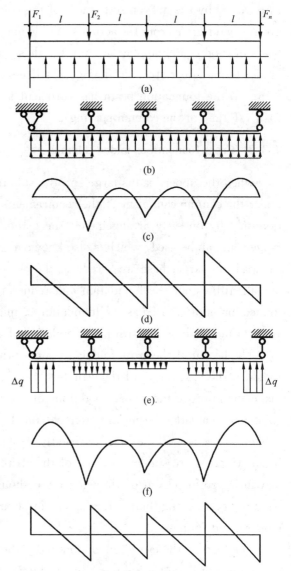

Figure 4-8　Calculation diagram of foundation beam inverted beam method

less than 20%, the adjustment calculation can be ended. Otherwise, the adjustment calculation is repeated until the accuracy requirements are met.

(6) Overlap the successive calculation results to obtain the final internal force distribution of the continuous beam, as shown in Figures 4-8(f) and(g).

Based on the linear distribution assumption of the subgrade reaction force, the inverted beam method calculates the subgrade reaction force according to the static equilibrium condition. This method regards the column end as a fixed support, ignoring the internal force caused by the overall bending of the beam and the secondary stress of the superstructure caused by the uneven settlement of the column foot. The calculated results are often obviously different from the actual situation and tend to be unsafe. Therefore, the inverted beam method can be applicable only for these situations: ① on a relatively uniform subgrade; ②the upper structure has good stiffness; ③the load distribution is relatively uniform; ④ the foundation beam has sufficient stiffness (the height of the beam is greater than 1/6 time of the column spacing).

4.3.3 Cross foundations under columns

when the upper load is large and the subgrade soil is weak, and the strip foundation under the column cannot meet the requirements of subgrade bearing capacity and foundation only by one-way arrangement, the orthogonal lattice foundation with two-way arrangement can be used, which is also known as cross foundation. Cross foundation spreads the load to a larger base area, reduces the pressure on the subgrade, and can improve the overall stiffness of the foundation and reduce the settlement difference. Therefore, this foundation is often used as the foundation of multi-storey buildings or high-rise buildings. And it can also be used with pile foundations for weak foundations.

The layout of the cross foundation under the column is shown in Figure 3-4. In order to adjust the load center of the structure to coincide with the center of the bottom surface plane and improve the stress conditions of the foundation under corner columns and side columns, foundation beams are often extended constructively at corners and side columns. The cross section of the beam is mostly T-shaped, and the design requirements of the beam structure are similar to those of the strip foundation. At the intersection, the two-way main reinforcement of the wing plate should be overlapped. If the foundation beam has torque, the longitudinal reinforcement should be configured according to the bending moment and torque.

The load on the cross foundation under the column is acted on the cross node by the column net through the column end, as shown in Figure 4-9. The basic principle of foundation calculation is to distribute the node load to the foundation beams in two directions, and then calculate the foundation beams in one direction by the above method respectively.

The distribution of nodal load on two orthogonal strip foundations must meet two conditions.

(1) The static equilibrium condition, that is, the sum of the loads distributed to the strip foundation in two directions at the node is equal to the column load, namely

$$F_i = F_{ix} + F_{iy} \tag{4-8}$$

where

F_i ——vertical column load on the node i, kN;

F_{ix} ——vertical load of x directional foundation beam at node i, kN;

F_{iy} ——vertical load of y directional foundation beam at node i, kN.

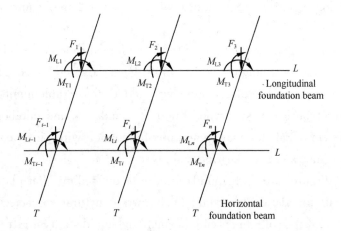

Figure 4-9 Stress diagram of cross foundation nodes

The bending moments at the node, M_x and M_y, are directly added to the foundation beam in the corresponding direction, so there is no need to redistribute. And the torque borne by the foundation beam is not considered.

(2) Deformation compatibility condition, that is, the vertical displacement of strip foundation in two directions at the intersection point should be equal after separation.

$$w_{ix} = w_{iy} \tag{4-9}$$

where

w_{ix} ——the vertical displacement of the x directional beam at the joint i;

w_{iy} ——the vertical displacement of y directional beam at the joint i.

According to Equations 4-8 and 4-9, each node can establish two equations. And only two unknowns F_{ix} of F_{iy}. The number of equations is the same as the number of the unknown quantity. So, all unknown quantities can be solved.

However, the actual calculation process is complicated, because the internal force and deflection of beams on elastic subgrade must be calculated using the above method to obtain the displacement of nodes. And the loads on these two groups of foundation beams are undetermined. That is to say, the distribution of column load and the internal force and deflection of two groups of elastic foundation beams must be solved jointly.

4.4 Raft foundations and box foundations

In the big cities, the land is very expensive, usually it is necessary to build large highrise buildings with a small floor area. This kind of building has a great load and has higher requirement for earthquake resistance. And some tall and complex structures are sensitive to subgrade settlement and uneven deformation. Obviously, it is difficult to meet the safety and use requirements by using pad foundation, strip foundation and pile foundation.

Chapter 4 Continuous Foundations

The raft foundations and box foundations which can meet the requirements are widely used.

Raft foundation is a whole continuous thick reinforced concrete slab buried in the subgrade. Box foundation is a single-layer or multi-layer box reinforced concrete structure consisting of bottom plate, top plate, outer wall and a considerable number of vertical and horizontal partitions buried in subgrade. Pile-raft foundation and pile-box foundation are the combined system of raft foundation and box foundation with piles penetrating through soft soil layer to dense and hard bearing soil layer

As shown in Figure 4-10, raft foundation can be divided into flat-plate raft foundation and beam-plate raft foundation according to its own structural characteristics. When the load is not large, and the column spacing is small and equidistant, a raft with equal thickness can be made, as shown in Figures 4-10(a) and (b). When the column load is large and uniform, and the column spacing is also large, in order to improve the bending stiffness of the raft, rib beams can be arranged along the vertical and horizontal axes of the column network to form a beam-slab raft foundation, as shown in Figures 4-10(c) and (d). When the ribbed beam is placed under the slab, the ground formwork method can be used to obtain a flat raft surface as the indoor floor. This is economical, but the construction is inconvenient, as shown in Figure 4-10(d). If the ribbed beam is set above the raft, the indoor floor will be overhead, but the column foundation can be strengthened and the construction is more convenient.

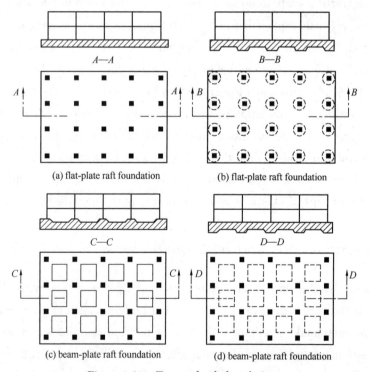

Figure 4-10 Types of raft foundations

4.4 Raft foundations and box foundations

When the subgrade is weak and uneven, buildings are sensitive to differential settlement, and the rigidity of raft foundation is not enough to adjust the possible differential settlement, box foundation can be used instead. Box foundation can be divided into single-layer box foundation (Figure 4-11a) and multi-layer box foundation (Figure 4-11b) according to its own stiffness characteristics.

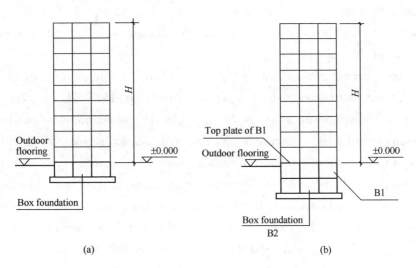

Figure 4-11 Single-layer and multi-layer box foundation

Compared with common foundations, raft foundations and box foundations have the following characteristics:

(1) The large foundation area can not only reduce the foundation pressure, but also improve the subgrade bearing capacity, so it is easy to bear the huge load of the superstructure and meet the requirements of the subgrade bearing capacity. Large foundation area can reduce the ratio of building height to foundation width and increase the stability of foundation. According to the statistics of raft foundation and box foundation at home and abroad, the height-width ratio of high-rise buildings is generally 6 : 1-8 : 1. At the same time, it should also be noted that the larger the foundation area, the deeper the additional stress of the foundation will spread. When there is a compressible soil layer buried deep in the foundation, it will cause greater foundation deformation.

(2) When the deep raft foundation and box foundation are used as the foundation of heavy buildings, they usually need to be buried to a certain depth, depending on the properties of the subgrade soil layer and the building. Generally, the minimum buried depth is 3-5 m. The foundation buried depth of modern high-rise and super-high-rise buildings has exceeded 20 m. For example, the buried depth of raft foundation of Tianjin 117 Building is 25 m. The box foundation of Beijing Capital Building is buried to a depth of 22.5 m. Because of the large buried depth, the additional stress of the foundation can be reduced due to the compensation effect, and the bearing capacity of the foundation can also be im-

proved, which is easy to meet the requirements of the bearing capacity of the foundation. Meanwhile, the foundation is deeply embedded in the subgrade, which is beneficial to reduce the subgrade deformation, increase the stability of the foundation and the seismic performance of the building. The Technical Specification for Raft-shaped and Box-shaped Foundations of High-rise Buildings stipulates that the buried depth of raft-shaped and box-shaped foundations on natural subgrade should not be less than 1/15 of the height of buildings in seismic areas. The buried depth of pile raft and pile box foundation should not be less than 1/18 of the building height.

(3) It has great rigidity and integrity, and can adjust the uneven deformation of subgrade through the mutual reaction of superstructure-subgrade-foundation. With the construction of high-weight multi-storey buildings, in order to distribute loads and to adapt to the continuous increase of column spacing, it is required to increase the stiffness of the foundation and make it into a thick raft foundation or a multi-storey box foundation. Now 3-5 m thick raft foundation and 3-4-storey box foundation are used in engineering.

(4) It can be combined with the basement construction. By increasing the buried depth of the foundation, the underground space above the foundation can provide good conditions to build the basement, and the raft has also become the floor of the basement.

(5) Raft and box foundations can be used in combination with pile foundations. When the buildings will be built on soft subgrade, and the uneven settlement should be strictly controlled, such as super-high-rise buildings on soft subgrade, only the bearing capacity requirements can be met by using raft or box foundations is used, while the settlement or settlement difference cannot be met. In this case, piling can be done under raft foundation or box foundation to reduce settlement, that is, pile raft foundation or pile box foundation.

(6) It is necessary to deal with the influence of large-scale deep excavation on the design and construction of raft foundation and box foundations. Large-scale deep excavation has to solve the problems of foundation pit slope support, artificial precipitation and its influence on adjacent buildings. Meanwhile, large-scale deep excavation has the most direct and important influence on foundation engineering such as the uplift of groundwater.

(7) Raft or box foundations are expensive and have technical difficulty because their construction will spend a lot of steel and concrete. Mass reinforced concrete construction needs to carefully control the quality and temperature. Meanwhile, large-scale deep excavation and deep burial of foundation have brought many geotechnical problems to be solved, which makes the cost much more expensive than that of common foundations.

Questions

4-1 What are the common properties of strip foundation under column, raft foundation and box foundation?

4-2 What is the Winkle subgrade model?

4-3 What is an elastic half space subgrade model? Explain the main differences between it and the Winkle subgrade model.

4-4 What is a finite compression layer subgrade model? What is the difference between it and the elastic half space subgrade model?

4-5 Summarize the advantages and disadvantages of the three foundation models mentioned above, and explain what conditions they are suitable for.

4-6 What are the basic assumptions of the inverted beam method? How to use the inverted beam method to calculate the internal force of the foundation beam?

4-7 How to distinguish short beams, finite beams and infinite beams when analyzing the internal forces of foundation beams using the Winkle foundation model?

4-8 What are the main characteristics of raft foundation and box foundation?

4-9 What are pile raft foundation and pile box foundation?

Chapter 5   Pile Foundations

5.1 Introduction

When the shallow foundation on natural subgrade can't meet the bearing capacity or settlement requirements of buildings, deep foundations can be used to transfer the load to deeper soil layers. Deep foundations include pile foundation, pier foundation and open caisson foundations.

The pile foundations refer to a rod-shaped member arranged vertically or slightly obliquely in the foundation, and its cross-sectional area is very small relative to its length. It transfers the load of the superstructure to the deep soil through the side friction and tip resistance of the bar. The pile foundation is an ancient foundation form. As early as prehistoric times, people used wooden stakes to cross river valleys and swamp areas. In the Hemudu site in Zhejiang province 7000 years ago, it was found that the ancients had used wooden stakes to support houses. Yuhe Bridge in Beijing, Longhua Tower in Shanghai and Baqiao Bridge in Xi'an are all examples of using wooden stakes in ancient China. In recent years, the use of piles has become more and more extensive, and the forms of piles have also developed greatly. Especially since 1980s, China's economic construction and civil engineering construction have developed rapidly, which makes great progress in pile technology. Tens of millions of piles have been used every year in China in the past twenty years.

Although the cost of pile foundation is generally higher than that of shallow foundation in natural subgrade, it can greatly improve the bearing capacity of foundation, reduce settlement, bear horizontal load and upward pull-out load, and has good seismic (vibration) performance, so it is widely used. At present, pile foundation is mainly used in the following aspects, which is shown in Figure 5-1.

(1) The load from the upper structures is very large, and the bearing stratum that can meet the requirements of bearing capacity can only be found in the deeper soil layers.

(2) In order to reduce the settlement or uneven settlement of the foundation, less piles are used to transfer part of the load to the depth of the foundation, thus reducing the settlement of the foundation. According to the settlement control design, this kind of pile foundation is called settlement reduction composite sparse pile foundation.

(3) When the bottom surface of the design foundation is higher than the natural

ground or the soil at the bottom of the foundation may be eroded, a pile foundation with high bearing platform is formed without contact between the bearing platform and the foundation soil.

(4) There are great horizontal loads, such as wind, waves, horizontal earth pressure, earthquake load and scour.

Vertical piles, inclined piles or cross piles can be used to bear horizontal loads.

(5) The underground water level is high, and deep foundation excavation and artificial dewatering are needed to deepen the foundation burial depth, which may be uneconomical or have adverse effects on the environment. At this time, pile foundation can be considered.

(6) Under the buoyancy of water, the basement or underground structure may float, and then the pile is used to resist floating and bear the uplift load.

(7) In the case of machine foundation, pile foundation can be used to control the amplitude and natural frequency of foundation system.

(8) Pile passing through collapsible soil, expansive soil, artificial fill, garbage soil, silt, swamp soil and liquefiable soil can ensure the stability of the building.

In addition to the use of pile foundation in the above situations, piles are also widely used as retaining structures of foundation pits, as well as piles as anchoring structures, and anti-slide piles for landslide control. Figure 5-1 shows several cases of using piles.

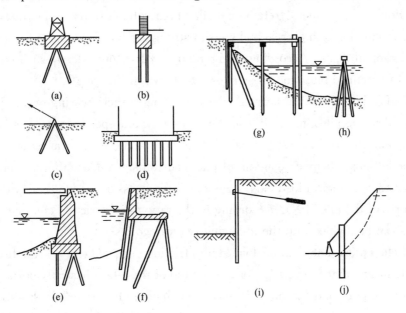

Figure 5-1　Engineering application of piles

This chapter focuses on the load transfer mechanism and design principles of piles. It mainly refers to *Code for Design of Building Foundation* (GB 50007—2011) and the *Technical Code for Building Pile Foundations* (JGJ 94—2008).

5.2 Classification and selection of piles

5.2.1 Classification of piles

1. Classification according to the function of piles

According to the use function of piles, they can be divided into the following four categories.

(1) Vertical compressive pile. This is the most widely used type of pile. Its pile foundation can improve the bearing capacity of foundation and (or) reduce the settlement of foundation.

(2) Vertical uplift piles such as anti-floating piles. With the construction of large-span light structures (such as airport apron) and shallow underground structures (such as underground parking lot), the use of such piles is becoming more and more extensive, and the amount is often large. The anchor pile used in the vertical static load test of single pile also bears the pull-out load.

(3) Horizontal loaded piles. They mainly bear horizontal loads, and the most typical ones are anti-slide piles and row piles in foundation pit retaining structure.

(4) Composite loaded piles. The vertical and horizontal loads of composite loaded piles are large. For example, piers, retaining walls, high-voltage transmission line towers and piles in the foundation of high-rise buildings in strong earthquake areas also bear large vertical and horizontal loads. According to the nature of horizontal load, such piles can also be designed as inclined piles and cross piles, as shown in Figure 5-1.

2. Classification of vertical bearing piles according to their bearing properties

Vertical bearing piles generally transfer loads to the deeper foundation soil below the pile cap through the friction resistance of the pile body and the end bearing force of the pile end. Figure 5-2 is a schematic diagram of pile load transfer under different conditions. In the figure, Q is the vertical load (kN), τ is the shear stress of the pile side (kPa) and N is the axial force of the pile (kN). According to the ratio of pile friction and pile end bearing force, piles can be divided into the following two categories.

(1) **Friction piles.** They can be divided into friction piles and end-bearing friction piles. Friction pile means that the vertical load at the top of the pile is basically borne by the side resistance of the pile, and the tip resistance is negligible (Figure 5-2a). End-bearing friction pile refers to the most commonly used pile, in which the vertical load at the top of the pile is mainly borne by the pile side resistance (Figure 5-2b).

(2) **End-bearing piles.** They can be divided into two types: end-bearing piles and friction end-bearing piles. End-bearing pile means that the vertical load at the top of the pile is basically borne by the resistance at the end of the pile, and the side resistance of the pile is

5.2 Classification and selection of piles

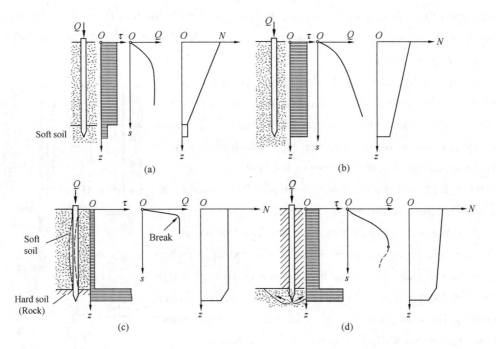

Figure 5-2 Schematic diagram of friction pile and end bearing pile

negligible (Figure 5-2c). Friction end bearing pile means that the load on the top of the pile is mainly borne by the resistance at the end of the pile under the limit state of bearing capacity (Figure 5-2d). The division of these four kinds of piles mainly depends on the distribution of soil layers, but it is also related to the conditions such as pile length, pile stiffness, pile shape, whether the bottom is enlarged or not, and pile forming method. For example, with the increase of pile length-diameter ratio, the load transmitted to the pile end will be reduced under the limit state of bearing capacity. And the play of side resistance $\frac{l}{d}$ and tip resistance at the lower part of the pile body will be relatively reduced. When $\frac{l}{d} \geqslant 40$, the load sharing ratio of tip resistance tends to zero in uniform soil layer; When $\frac{l}{d} \geqslant 100$ and the pile end is located on the hard soil (rock) layer, the load sharing value of the tip resistance is too small such that it can be ignored.

3. Classification according to the pile materials

Wood pile is the oldest pile material, but it is rarely used at present because of the limitation of resources, easy corrosion and difficult to extend. Therefore, modern piles can be divided into the following three categories according to materials.

(1) Concrete piles. They are generally made of reinforced concrete. According to the construction method, it can be divided into cast-in-place pile and precast pile. Precast piles can be divided into two types: site prefabrication and factory prefabrication, and the latter

is subject to the test of transportation. Precast piles can also be divided into prestressed piles and non-prestressed piles. Prestressed piles made of high-strength cement and steel bars have high pile strength.

(2) Steel piles. They can be divided into steel pipe piles, steel sheet piles, steel piles and composite section piles according to the section shape, as shown in Figure 5-3. Steel piles are easy to be driven into the soil, which causes little disturbance to the stratum because of less soil squeezing, but the cost is high and the corrosion resistance is poor, so surface anti-corrosion treatment is needed.

(3) Composite piles. There are many kinds of composite piles, and new types appear constantly. For example, when used as an anti-slide pile, large I-beams are added to concrete to bear horizontal loads. In the cement wall made by deep mixing method, H shaped steel is inserted to form underground continuous wall. Recently, in China, researchers inserted high-strength reinforced concrete piles into cement soil as the core, and the bearing capacity of the piles formed was higher than that of ordinary cast-in-place piles. Another kind of rammed expanded pile with composite carrier is to ram masonry at the end of the pile, ram hard concrete on it, and then pour reinforced concrete pile, which is widely used.

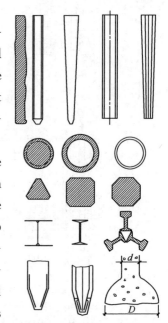

Figure 5-3　Piles with different cross-sections

4. Classification by pile-forming method

According to the method of pile formation, there are also various criteria for classifying piles. Among them, piles are often classified into three categories based on whether or not the soil isdisplaced during the pile formation process, which has a significant impact on the properties of the pile and foundation soil.

(1) Non-displacement piles. Non-displacement piles are characterized by taking soil in advance to form holes, and the method of forming holes is to drill holes with various rigs or dig holes manually. Figure 5-4 shows typical construction steps of non-displacement piles. Manual digging is usually above the groundwater level Drilling holes can be on water or under water. When drilling underwater, it is necessary to protect the hole wall. Usually, slurry is used to protect the drilled hole wall, that is, slurry is injected into the hole, and the slurry level is kept 1-2 m higher than the groundwater level to ensure the stability of the drilled hole wall.

(2) Displacement piles. Soil compaction piles are mainly precast piles. Construction method: the precast pile is implanted into the foundation soil by hammering, vibration or static pressure, so that the foundation soil occupied by the pile body is squeezed around the

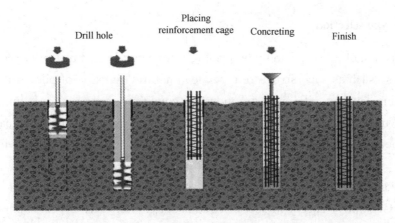

Figure 5-4 Construction steps of non-displacement piles

pile. In a suitable soil layer (such as a compacted soil layer with low saturation), a closed casing with a movable flap door at the bottom of the pipe can also be driven into the foundation soil, and concrete is poured after the pipe is pulled out into the hole that is formed. The pile formed in this way is called a soil-squeezing cast-in-place pile. For saturated soft clay, when there are many compacted piles sinking, because the soil is incompressible in a short time, it may lift the ground, cause damage to adjacent buildings or pipelines, and cause the buried piles to float, move sideways or break. At the same time, it will cause higher excess pore water pressure in foundation soil, which is very unfavorable.

(3) Partial displacement piles. Open-ended casing soil extraction and pile pouring, first involves pre-drilling smaller diameter boreholes, and then inserting prefabricated piles. Driven open-ended pipe piles and similar methods are all classified as partial displacement piles.

5. Classification according to the geometric characteristics of piles

The geometric size and shape of piles are very different, which has a great influence on the bearing behavior of piles. In this respect, piles can also be classified from different angles. According to the different pilediameters d, piles can be divided into the following three categories:

(1) large diameter piles: $d \geqslant 800$ mm;

(2) medium diameter piles: $250 \text{ mm} < d < 800$ mm;

(3) small diameter piles: $d \leqslant 250$ mm.

It is generally believed that for cast-in-place piles with a diameter greater than 800 mm, the bearing capacity may be reduced due to the relaxation of soil stress around the pile hole, especially for sandy soil and gravel soil. In this case, it should multiply the size effect coefficient less than 1.0. Small diameter piles with small pile diameter and large slenderness ratio are also called micropiles.

5.2.2 Pile type selection

The selection of pile type and pile-forming technology should be made according to local conditions, such as superstructure type, load nature, function of pile, soil type in soil layers, bearing layer at pile end, groundwater condition, construction equipment, construction environment, construction experience and supply of pile-making materials. Its principles should be economic and reasonable, safe and applicable, and environmentally friendly.

Table 5-1 is a reference table for the selection principle of main types of piles in pile foundation. Generally, except in special circumstances, different types of piles should be avoided in the same building.

Selection of pile type Table 5-1

Pile type	Building type	Stratigraphic conditions	Construction conditions
Precast pile	General high-rise and multi-storey buildings; Industrial and civil buildings and structures with high requirements for foundation settlement	Uneven surface soil and thickness. The groundwater level is shallow, and there is possibility of shrinkage cavity. There is a good bearing stratum available within a certain depth; There is no impenetrable hard interlayer on the upper part and no soil layer sensitive to soil squeezing effect	The site is empty, adjacent to no dangerous buildings, and there are no restrictions on noise, vibration and lateral extrusion
Cast-in-place pile	General high-rise residential buildings and multi-storey buildings	The available bearing layer at the pile tip fluctuates greatly or there is a hard interlayer above the bearing layer that is not easy for precast piles to penetrate, and there is no shrinkage cavity phenomenon	(1) Requires a certain site, for construction machinery loading and unloading and transportation; (2) Can solve the problem of unearthed stacking during construction; (3) There are no obstacles underground
Short pile and short pile with enlarged bottom	Generally, buildings with less than 6 floors	The topsoil is poor, the thickness of the filled soil is below 4-6m, and there is general Quaternary soil available, while the hard layer and groundwater level are deep	(1) Requires a certain site, for construction machinery loading and unloading and transportation; (2) Can solve the problem of unearthed stacking during construction; (3) There are no obstacles underground
Large diameter pile	Important large public buildings or high-rise residential buildings, industrial and civil buildings and structures with strict requirements for foundation settlement	The surface soil quality and thickness are uneven, there is no shrinkage cavity, and there is a good bearing soil layer within a certain depth	If mechanical drilling is used, a certain site is required for loading, unloading and transportation of drilling machinery; If artificial pore-forming is used, adequate safety and quality assurance measures should be taken

5.3 Load transfer of vertical bearing piles

The main function of pile foundation is to transfer vertical load to the lower soil layer, which is carried out through the interaction between pile and its periphery soils.

5.3.1 Load transfer mechanism

The vertical pressure acting on the pile top is beared by the total frictional resistance acting on the pile side Q_s and the total tip resistance acting on the pile end Q_p, as shown in Figure 5-5, which can be expressed as

$$Q = Q_s + Q_p \qquad (5\text{-}1)$$

The exertion process of pile side resistance and pile tip resistance is the load transfer process of pile-soil system. After the top of the pile is subjected to vertical pressure, the pile body is compressed and displaced downwards relative to the adjacent soil. As a result, the side surface of the pile is subjected to upward frictional resistance of the soil, and the load is transferred to the soil around

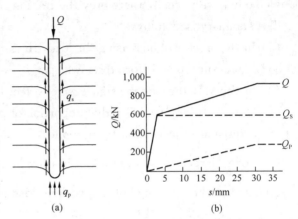

Figure 5-5 Side resistance and tip resistance of pile

the pile through the lateral resistance. It can be seen that the axial force and compressive deformation of the pile body decrease with the depth. With the increase of load, the lateral resistance of the lower part of the pile gradually plays a role. When the load reaches a certain value, the vertical displacement of the pile end begins and the reaction force at the pile end begins to play a role. Therefore, the lateral resistance of the upper soil layer near the pile body acts earlier than that of the lower soil layer.

When getting displaced, the side resistance plays a role before the tip resistance. The research shows that the displacement required by the side resistance and the tip resistance is also different. The test results of a large number of conventional diameter piles show that the relative displacement required for the lateral resistance to play a role is generally less than 20 mm. For large-diameter piles, generally under the condition of displacement $s=(3\%\text{-}6\%)d$, lateral resistance has played most of the role. However, the effect of tip resistance is more complicated, which is related to the type and nature of soil at the end of pile, pile length, pile diameter, pile forming technology and construction quality.

For rock stratum and hard soil layer, only a small displacement of pile tip can fully exert its tip resistance, while for general soil layer, the displacement required to fully exert its tip resistance may be large. Taking the bearing layer at the pile end as fine-grained soil as an example, to give full play to the role of tip resistance, the $\frac{s_p}{d}$ (the ratio of the settlement at the pile end to the pile diameter) of the driven pile should be about 10% and the $\frac{s_p}{d}$ of bored piles can reach 20%-30%.

In this way, for the general pile foundation, under a working load, the side resistance plays most of the role, while the tip resistance only plays a small part. Only the rigid short pile supported on hard rock foundation has a tip resistance that plays a role before the side resistance, this is because its pile end is difficult to sink, and the compression of the pile body is very small, therefore, the friction resistance cannot play a larger role.

The summary is as follows.

(1) In the process of increasing the load, the lateral resistance of the upper part of the pile body comes into play before the lower part.

(2) In general, the side resistance takes effect before the tip resistance.

(3) Under working load Q_k, the proportion of side resistance is obviously higher than that of tip resistance for general friction piles.

(4) For $\frac{l}{d}$ larger piles, even if the bearing layer at the pile end is rock or hard soil, the resistance at the lower end of the working load is very minimal due to the compression of the pile itself. When $\frac{l}{d} \geqslant 100$, the tip resistance can be basically ignored and become a friction pile.

Figure 5-6 shows the effect of lower tip resistance and side resistance in three cases. Q_k is the corresponding standard combination of a single pile's vertical force, Q_u corresponds to the ultimate load of the single pile, Q_{su} is the total side resistance under the ultimate load and Q_{pu} is the total tip resistance under the ultimate load.

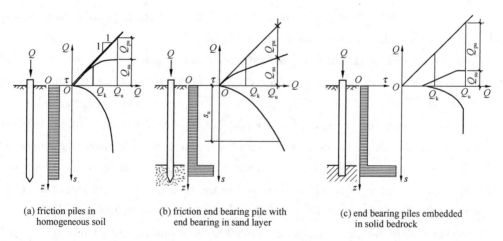

(a) friction piles in homogeneous soil
(b) friction end bearing pile with end bearing in sand layer
(c) end bearing piles embedded in solid bedrock

Figure 5-6 Tip resistance and side resistance under several conditions

5.3.2 Side resistance of piles

As mentioned above, the effect of pile side friction is related to the relative displacement between the pile and soil. For friction piles, when there is vertical pressure Q on the top of the pile, the displacement of the top of the pile is s_0. s_0 consists of two parts: one part is the settlement of the pile end s_p, which includes the compression of the soil at the pile end and the displacement of the whole pile body caused by the penetration of the pile tip into the soil at the pile end. The other part is the compression deformation of the pile under the action of axial force s_s. As Figure 5-7(e) shown, $s_0 = s_p + s_s$. If the single pile shown in Figure 5-7 has a length of l, a cross-sectional area of A, a diameter of d and the elastic modulus of the pile material is E, then the measured distribution curve of axial force $N(z)$ of each section along the buried depth of the pile is as shown in Figure 5-7(c). It can be seen that the axial force $N(z)$ decreases with the increase of depth z. The decreasing rate reflects the amount of unit side resistance q_s. In Figure 5-7(a), the micro-section dz of the pile taken at the depth z can be obtained according to the balance condition of the vertical force of the micro-section (ignoring the self-weight of the pile).

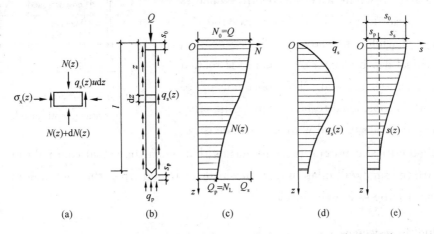

Figure 5-7 Distribution of axial force, side friction and settlement pile

$$q_s(z)\pi d\,dz + N(z) + dN(z) - N(z) = 0$$

$$q_s(z) = \frac{1}{\pi d}\frac{dN}{dz} \tag{5-2}$$

Equation 5-2 shows that at any depth z, the unit lateral resistance q_s due to the relative displacement s between pile and soil is directly proportional to the change rate of axial force N of pile at this place. Equation 5-2 is called the basic differential equation of pile load transfer.

After measuring the vertical displacement of the pile top s_0, the displacement of the pile end and the displacement of the pile section at any depth $s(z)$ can also be calculated by using the above-mentioned measured axial force $N(z)$ distribution curve, that is

$$s_p = s_0 - \frac{1}{AE}\int_0^l N(z)\,dz \qquad (5\text{-}3)$$

$$s(z) = s_0 - \frac{1}{AE}\int_0^z N(z)\,dz \qquad (5\text{-}4)$$

Figure 5-7(e) shows the vertical settlement distribution of the pile. It is worth pointing out that the load transfer curve ($N-z$ curve), the side resistance distribution curve q_s-z curve) and the vertical displacement curve ($s-z$ curve) of the pile in Figure 5-7 are constantly changing with the increase of the pile top load.

Many measured load transfer curves show that the distribution of q_s may be in various forms. For driven piles, in cohesive soil, the distribution of q_s along the depth is similar to a parabola, as shown in Figure 5-7d. Under the ultimate load, the median value q_s of sand increases approximately linearly with the depth at first, and then nearly uniformly distributes after reaching a certain depth, which is called the critical depth. Figure 5-8 shows the test results of model piles in sandy soil. It can be seen that the critical depth of lateral resistance is related to the compactness of sand. This phenomenon of critical depth is considered to be related to the critical confining pressure of sand with a certain density.

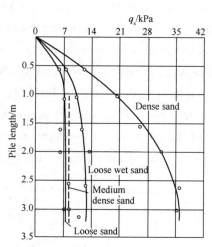

Figure 5-8 Side resistance test and critical depth of pile

The so-called critical confining pressure means that when the actual confining pressure is less than it, the sand will dilate under the shear load, and when the confining pressure is greater than it, the sand will shrink.

5.3.3 Pile tip resistance

The pile tip resistance is an important part of its bearing capacity, and it is influenced by many factors. The function of the pile tip resistance is related to types of piles and soil.

Before the 1960s, people mostly used the classical bearing capacity theory based on the assumption that the soil is rigid and plastic to analyze the pile tip resistance. The pile is considered as a foundation with a width b(equivalent to the diameter of the pile d) and a buried depth l of the pile. When the pile is loaded, the soil at the pile tip is sheared. According to the assumed different slip surface shapes, the ultimate bearing capacity of the pile tip is calculated by using the theory of ultimate bearing capacity of foundation introduced in the soil mechanics textbook, and the ultimate unit tip resistance q_{pu} is determined. However, because the depth of the pile is much larger than the cross-section size of the pile, most of the soil at the pile end belongs to impact shear failure or local shear failure. Only when the pile length is relatively short and the pile passes through the soft soil layer and is supported on the solid soil layer, the overall shear failure similar to the foundation under shallow foundation may occur. Figure 5-9 shows the commonly used sliding surface shapes of Karl Terzaghi type and Meyerhof type. According to the ultimate bearing capacity theory, the general expression of q_{pu} is:

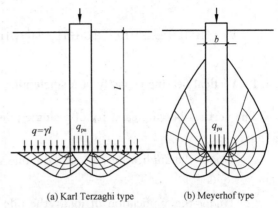

Figure 5-9 Two modes of foundation failure at pile tip

$$q_{pu} = \frac{1}{2} b\gamma N_\gamma + cN_c + qN_q \tag{5-5}$$

where

N_γ, N_c, N_q ——bearing capacity coefficient, whose value is related to the internal friction angle φ of soil, please refer to the soil mechanics textbook;

$b(d)$ ——width or diameter of pile, mm;

c ——cohesion of soil, kPa;

q ——vertical self-weight stress in soil at the elevation of pile bottom, $q = \gamma l$, kPa.

The tip resistance of pile, like the bearing capacity of shallow foundation, also mainly depends on the type and nature of soil at the end of pile. Generally speaking, coarse-grained soil is higher than fine-grained soil. Dense soil is higher than loose soil.

The tip resistance of the pile is greatly influenced by the pile-forming technology. For displacement piles, if the soil around the pile is compacted (such as loose sand), the soil at the end of the pile will be compacted to increase the tip resistance, and the tip resistance will play a role under the small displacement of the pile end. For dense soil or saturated cohesive soil, the result of compaction may not be causing more soil compaction, but disturb the structure of undisturbed soil, or produce excess pore water pressure, and the tip resistance may be adversely affected. For non-displacement piles, undisturbed soil may be disturbed when the pile is formed, causing dreg and loosened soil to be formed at the bottom of the pile, so the tip resistance will be obviously reduced. Among them, the tip resistance of large diameter bored pile decreases with the increase of pile diameter due to stress relaxation caused by excavation.

For the underwater cast-in-place pile, the tip resistance is generally smaller than that of dry cast-in-place pile because the sediment at the bottom of the pile is not easy to clean.

5.4 Vertical bearing capacity of piles

5.4.1 Vertical bearing capacity of a single pile

The vertical bearing capacity of a single pile should meet the following three requirements.

(1) Under the application of load, the pile will not lose its stability in the foundation soil.

(2) Under the application of load, the pile top does not produce excessive displacement.

(3) Under the application of load, the pile material will not be damaged.

There are many factors affecting the vertical bearing capacity of a single pile, such as soil type, soil properties, pile material, pile diameter, pile penetration depth and construction technology. In the long-term engineering practice, people have put forward many methods to determine the bearing capacity of a single pile. At present, the following methods are mainly adopted in engineering practice.

1. In-situ vertical static load test

The vertical static load test of a single pile can be carried out before the pile construction to determine the bearing capacity of a single pile. It can also be used to detect the engineering piles after construction. The pile is formed in the construction site according to the design and construction conditions. The material, length, section and construction method of the test pile are consistent with the actual engineering pile. It is suitable for bearing capacity of single pile under various conditions, especially for these cases when important buildings or geological conditions are complex, the reliability of pile construction quality is low and it is difficult to accurately determine the vertical bearing capacity of single pile by other methods. The number of test piles under the same conditions should not be less than 1% of the total number of piles, and should not be less than three. Figure 5-10 is the schematic diagram of two kinds of vertical static load tests of a single pile commonly used in engineering. During the test, jack is used to load the pile top step by step, and the settlement s of the pile top under each load is recorded when the deformation is stable. The load Q and settlement s curve of pile top drawn by the test results are shown in Figure 5-11.

According to the measured curve, the ultimate vertical bearing capacity of a single pile can be determined by the following methods.

(1) When the steep drop section of the curve is obvious, take the load value of the starting point of the corresponding steep drop section, i.e point B as shown in Figure 5-11.

(2) When the curve is slowly changing, take the load value corresponding to the total

5.4 Vertical bearing capacity of piles

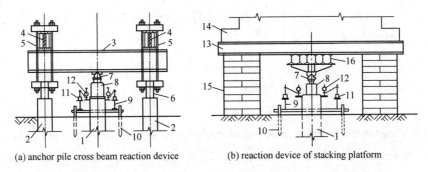

(a) anchor pile cross beam reaction device (b) reaction device of stacking platform

Figure 5-10 Two kinds of static load test devices for a singlep pile

1-test pile; 2-anchor pile; 3-main beam; 4-secondary beam; 5-pull rod; 6-Anchor bar;
7-tee; 8-Jack; 9-a benchmark beam; 10-benchmark pile; 11-magnetic watch base;
12-displacement meter; 13-load platform; 14-stacking; 15-buttress; 16- joist

settlement of the pile top $s=40$ mm; when the pile length is more than 40 m, elastic compression of the pile body can be considered, and the corresponding total settlement can be appropriately increased.

(3) When $\dfrac{\Delta s_{n+1}}{\Delta s_n} \geqslant 2$ appears in the test and within 24 hours stability is not achieved, take the load corresponding to s_n. In the equations: $\Delta s_n = s_n - s_{n-1}$, $\Delta s_{n+1} = s_{n+1} - s_n$, which are the settlement increment of pile top caused by the n level and $n+1$ level loads respectively.

(4) If it is difficult to judge by the above methods, it can be comprehensively judged by other auxiliary methods. If there are special requirements for foundation settlement, it can be selected according to specific conditions.

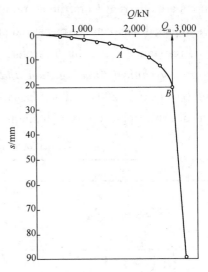

Figure 5-11 Q-s curve of single pile test

For the test piles participating in statistics, when the range of vertical ultimate bearing capacity of each single pile does not exceed 30% of the average value, the average value can be taken as the standard value of vertical ultimate bearing capacity of a single pile. When the range exceeds 30% of the average value, it is advisable to analyze the reasons for this and increase the number of test piles, then determine the ultimate bearing capacity according to the specific conditions of the project. For the case that the pile cap under the column has only three piles or less, the minimum value is taken.

when the standard value of ultimate bearing capacity of a single pile, determined above, is divided by a safety factor K, the result gives a single pile's characteristic value of vertical bearing capacity of R_a as shown in Equation 5-6.

$$R_a = \frac{1}{K}Q_{uk} \tag{5-6}$$

where

Q_{uk} ——standard value of vertical ultimate bearing capacity of a single pile;

K ——safety factor, take $K=2.0$.

2. Cone Penetration Test (CPT) method

For buildings with foundation design Grade C, in-situ static cone penetration test can be used to determine the characteristic value of vertical bearing capacity R_a of a single pile.

Static cone penetration test is very similar to the process of pile entering the soil, so it can be regarded as a field simulation test of small-sized hitting piles. Because of its simple equipment and high degree of automation, it is considered to be a promising method to determine the bearing capacity of single pile. However, because the dimensions and conditions are different from the static load test of piles, an empirical relationship is generally established between the measured specific penetration resistance p_s, side resistance q_{sk} and end resistance q_{pk}. Table 5-2 refers to the empirical relationship recommended by the *Code for Geotechnical Investigation and Design of Building Foundations in Beijing Area* (*DBJ 11-501—2009*). Then Equation 5-7 can be used to determine the standard value of vertical bearing capacity of a single pile.

Estimation of q_{sk} and q_{pk} according to p_s of CPT test Table 5-2

Name of soil	p_s/MPa	q_{sk}/kPa	q_{pk}/kPa
Cohesive soil	0.5-1.0	30-40	—
	1.0-1.5	40-50	—
	1.5-2.0	50-60	—
	2.0-3.0	60-70	—
Silt	1.0-3.0	40-70	2,000-3,000
	3.0-6.0	70-90	3,000-4,000
Sand	5.0-15.0	40-60	3,600-4,400
	15.0-25.0	60-80	4,400-6,400
	25.0-30.0	80-90	6,400-8,400

Note: The specifications of the probe for CPT are cone bottom area of 10 cm², cone angle of 60, sidewall length of 7 cm, and penetration speed of 0.8-1.4 m/min.

3. Empirical method

The standard value of ultimate vertical bearing capacity Q_{uk} of a single pile can be estimated by Equation 5-7:

$$Q_{uk} = Q_{sk} + Q_{pk} = u\sum q_{sik} l_i + q_{pk} A_p \tag{5-7}$$

where

u ——circumference of pile;

l_i ——thickness of the pile passing through the ith layer of rock and soil;

A_p —— cross-sectional area of pile bottom;

q_{sik} —— standard value of ultimate lateral resistance of the ith layer of soil at the pile side;

q_{pk} —— standard value of ultimate end resistance.

From above, the q_{sik} and q_{pk} can be obtained according to the statistical analysis of static load test results in different areas. When there is no local experience, it can also be taken according to the physical properties of soil.

When using Equation 5-7 to estimate the standard value of vertical ultimate bearing capacity of single pile, the following points should be noted.

(1) For large-diameter bored piles ($d>800$ mm), q_{sik} and q_{pk} should be multiplied by the size effect coefficient less than 1.0.

(2) For rock-socketed piles, the standard value of total ultimate resistance of rock-socketed section can be determined according to the standard value of saturated uniaxial compressive strength of rock.

(3) For post-grouting cast-in-place piles, the ultimate side resistance and end resistance are multiplied by the enhancement coefficient greater than 1.0.

5.4.2 Pile group effect

A pile foundation consisting of three or more piles is called group pile foundation. Due to the interaction among piles, soil between piles and pile caps, the bearing capacity and settlement behavior of pile groups are obviously different from that of a single pile. After the pile group foundation is stressed (mainly under vertical pressure), its total bearing capacity is often not equal to the sum of the bearing capacity of each single pile, which is called pile group effect. The pile group effect not only occurs under vertical pressure, but also has shielding effect on the lateral bearing capacity of the rear piles when subjected to lateral force. When subjected to pullout force, the possible overall pullout of pile groups belongs to pile group effect. This part focuses on the analysis of pile group effect under vertical pressure.

Firstly, the interaction between pile and soil is analyzed. As mentioned above, for soil compaction piles, in sandy soil and silty soil with low saturation and general cohesive soil, the soil is compacted due to the soil compaction effect of pile groups, thus increasing the lateral resistance of piles. However, sinking more compaction piles in saturated soft clay will cause excess pore water pressure, thus reducing the bearing capacity of piles. And negative friction will occur with the consolidation and settlement of foundation soil.

The force borne by the pile will eventually be transferred to the foundation soil. For end-bearing piles, the force on the pile is directly transmitted to the soil layer at the pileend through the pile body. If the soil layer is hard and the area under pressure at the pile end is very small, the pressures at the pile ends will basically not affect each other, as shown in Figure 5-12(a). In this case, the settlement of pile group is basically the same as that of a single pile, and the bearing capacity of pile group is equal to the sum of the bearing capacity of each single pile. Friction pile transfers the vertical force to the soil around the pile through the friction force on the side of the pile, and then to the soil at the pile end. It is generally believed that the vertical additional stress caused by pile side friction in the soil spreads down to the pile end plane at a certain angle θ, as shown by the shadow in Figure 5-12. When the number of piles is small and the pile spacing s_a is large, for example $s_a > 6d$ (d is pile diameter), the additional pressure from each pile at the pile end plane does not overlap each other or does not overlap much (Figure 5-12b), then the working state of each pile in the pile group is similar to that of a single pile. However, when the number of piles is large and the pile spacing is small, such as the common pile spacing $s_a = (3\text{-}4)d$, the pressure from each pile in the foundation at the pile end will be superimposed on each other (Figure 5-12c), which makes the pressure at the pile end increase compared with that at the single pile. Meanwhile, the load area is widened, and the influence depth is deeper. As a result, on one hand, the total pressure of the bearing layer at the pile tip may exceed the bearing capacity of the soil layer; on the other hand, the settlement of pile group foundation is much higher than that of single pile due to the increase of additional stress, and the range is widened and deepened, especially if there is a highly compressible soil layer under the bearing layer at the pile end, as shown in Figure 5-12(d), the bearing capacity of pile may be obviously reduced due to settlement control. For end-bearing friction piles and friction end-bearing piles, due to the diffusion of friction force of pile groups and the end-bearing pressure of adjacent piles, the additional stress on the outer side of the bottom surface of each pile end increases, which is equivalent to increasing the vertical self-weight stress q in the Equation 5-6 for calculating the end-bearing capacity of piles, which will also improve the end-bearing capacity of single piles.

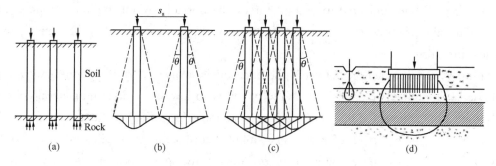

Figure 5-12 Pile group effect

Secondly, the cap also plays an important role in the pile group effect. The bearing platform is in direct contact with the soil between piles under the bearing platform, and the bearing platform will move downward under the vertical pressure, and the soil surface between piles will bear pressure, which will share the load acting on the piles, and sometimes the load will be as high as 1/3 or even higher than the total load. Only in the following cases, the pile cap and the soil surface may be separated or not in close contact, resulting in the non-existence or unreliability of load sharing: 1) The pile foundation bears the frequent dynamic action, such as the pile foundation of the railway bridge. 2) There are soil layers that may generate negative friction under the bearing platform, such as collapsible loess, unconsolidated soil, newly filled soil, high-sensitivity clay and liquefiable soil. 3) Dense pile groups are sunk in saturated soft clay, which causes excess pore water pressure and soil uplift, and then the soil between piles gradually consolidates and sinks. 4) Stacking load or precipitation around the pile may cause the ground around the pile to be separated from the pile cap, etc. However, in the design, for safety reasons, the bearing effect of soil between piles under the bearing platform is generally ignored.

Thirdly, the pile cap also has an influence on the friction and end bearing capacity of each pile. Because the soil, pile and pile cap have basically the same displacement at the bottom of pile cap, the relative displacement between this part of pile and soil is reduced, so that the pile side resistance at the top of pile cannot be fully exerted. In another aspect, the vertical additional stress exerted on the ground by the bottom surface of the pile cap increases the side resistance and end resistance of the pile. Rigid pile caps are used to connect the pile groups, which can adjust the stress of each pile. Although the vertical displacement of each pile top is basically equal under the central load, the vertical force shared by each pile is not equal. Generally, the stress distribution of corner piles is greater than that of side piles, and that of side piles is greater than that of central piles, that is, saddle distribution. At the same time, the overall effect will make the piles with good quality and high stiffness bear more stress, while the piles with poor quality and low stiffness bear less stress, and finally make the piles work together which increases the overall reliability of the pile foundation.

In a word, some pile group effects are favorable and some are unfavorable, which are related to many factors, such as the distribution of soil layers of pile group foundation, the nature of each soil layer, the pile spacing, the number of piles, the length-diameter ratio of piles, the ratio of pile length to cap width, and the pile-forming technology.

5.4.3 Vertical bearing capacity of a composite foundation pile

In the *Technical Code for Building Pile Foundations* (*JGJ* 94—2008), a composite pile foundation is defined as the pile foundation which bears the load by all piles and the subgrade soil under the pile cap. Each single pile in a composite pile foundation is called a composite foundation pile. If the bearing function of the subgrade soil under the pile cap is not considered, each single pile in pile foundation is called a foundation pile.

For end bearing pile foundations, independent friction type pile foundations with fewer than four piles, orwithout considering pile capping effects due to factors such as soil properties and usage conditions, the characteristic value of vertical bearing capacity of a foundation pile R should be taken as the characteristic value of vertical bearing capacity of a single pile R_a, that is $R=R_a$.

For friction type pile foundations that meet one of the following conditions, it is advisable to consider the pile capping effect to determine the characteristic value of the vertical bearing capacity of a composite foundation pile:

① buildings (structures) whose upper structure has good overall stiffness and simple shape;

② bent structures and flexible structures with strong adaptability to differential settlement;

③ the relatively stiffness weakened area of pile foundation which has optimized design of pile foundation stiffness to reduce differential settlement;

④ settlement reduction composite piles in soft soil subgrade.

The characteristic value of vertical bearing capacity of a composite pile foundation considering the pile capping effect can be determined according to the following equations:

without seismic effects:

$$R = R_a + \eta_c f_{ak} A_c \tag{5-8}$$

considering seismic effects:

$$R = R_a + \frac{\zeta_a}{1.25} \eta_c f_{ak} A_c \tag{5-9}$$

$$A_c = (A - nA_{ps})/n \tag{5-10}$$

where

η_c ——coefficient of pile capping effect, which can be taken as shown in Table 5-3;

f_{ak} ——the weighted average of the characteristic values of the subgrade bearing capacity of soil layers within a depth range of 1/2 of the pile cap width and not exceeding 5 m, kPa;

A_c ——net subgrade soil area under the pile cap corresponding to a composite foundation pile, m²;

A_{ps} ——cross-sectional area of the pile body, m²;

A ——the calculation area of the bearing platform. For independent pile foundations under the column, A is the total area of the bearing platform. For pile raft foundation, A is the area enclosed by the 1/2 span of the column, wall raft, and 2.5 times the raft thickness of the cantilever edge. Piles are concentrated under a single wall, and the area enclosed by 1/2 span on both sides of the wall is calculated as a strip bearing platform, m²;

ζ_a ——adjustment coefficient of seismic bearing capacity of subgrade soil according to Table 9-8.

5.5 Vertical bearing capacity verification of pile foundations

Coefficient of pile capping effect η_c   Table 5-3

S_a/d \ B_c/l	3	4	5	6	>6
⩽0.4	0.06-0.08	0.14-0.17	0.22-0.26	0.32-0.38	
0.4-0.8	0.08-0.10	0.17-0.20	0.26-0.30	0.38-0.44	
>0.8	0.10-0.12	0.20-0.22	0.30-0.34	0.44-0.50	0.50-0.80
Single row pile strip pile cap	0.15-0.18	0.25-0.30	0.38-0.45	0.50-0.60	

Note: 1. The S_a/d is the ratio of pile spacing to pile diameter. The B_c/l is the ratio of the pile cap width to the length of the pile. When the piles are not arranged in a square arrangement, take $s_a = \sqrt{A/n}$ in which A is the area of pile cap and n is the total number of piles.
2. For piles arranged under box and raft foundations, the η_c value can be taken as that of single row pile strip pile cap.
3. For single row pile strip pile caps, when the width of the pile cap is less than $1.5d$, the value shall be taken as that of non-strip pile caps.
4. For pile caps with post grouting piles, it is advisable to take a lower value.
5. For squeezed pile foundations in saturated cohesive soil and pile caps on soft soil foundations, it is advisable to take 0.8 times the lower value.

when the bottom of the pile cap is composed of liquefiable soil, collapsible soil, highly sensitive soft soil, under-consolidated soil, newly filled soil, or excessive pore water pressure and soil uplift produced due to pile driving, the pile capping effect should not be considered and $\eta_c = 1.0$ should be taken.

5.5 Vertical bearing capacity verification of pile foundations

5.5.1 Load applied on a foundation pile

Under the application of the load, the force bored by each pile in the pile group foundation is generally uneven and complicated, which is influenced by many factors. However, in practical engineering design, the vertical load is usually assumed to be linearly distributed. In this way, under the central vertical force, each pile bears its average value. Under the action of eccentric vertical force, the vertical force distributed on each pile changes linearly according to the distance from the centroid of the pile group, which is shown as Figure 5-13 and in the following equations:

under the action of central vertical force F_k,

$$N_k = \frac{F_k + G_k}{n} \quad (5-11)$$

under the action of eccentric vertical force F_k, M_{xk} and M_{yk},

$$N_{ik} = \frac{F_k + G_k}{n} \pm \frac{M_{xk} y_i}{\sum_{j=1}^{n} y_j^2} \pm \frac{M_{yk} x_i}{\sum_{j=1}^{n} x_j^2} \qquad (5\text{-}12)$$

where

N_k ——the vertical force on each pile under the central vertical force, kN;

n ——number of piles in the pile foundation;

N_{ik} ——vertical force on the ith pile under the action of eccentric vertical force, kN;

M_{xk}, M_{yk} ——moment acting on the bottom surface of the pile cap;

x_i, y_i ——distance from the center of the ith pile to the y, x axis.

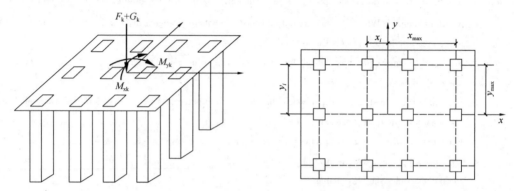

Figure 5-13 Load calculation applied on a foundation pile

5.5.2 Vertical bearing capacity verification of pile foundations

After determining the load on each pile in the pile foundation, verify the bearing capacity of a foundation pile using the following equations:

under the action of central vertical force,

$$N_k \leqslant R \qquad (5\text{-}13)$$

under the action of eccentric vertical force,

$$N_{max} \leqslant 1.2R \qquad (5\text{-}14)$$

For vertical compression piles, the bearing capacity characteristic values of a foundation pile should also consider the requirements of pile strength in design. Generally speaking, the bearing capacity of piles mainly depends on the supporting capacity of subgrade rock and soil. However, for end-bearing piles, super-long piles or defective piles, it may be controlled by the concrete strength of pile body. Due to the design related to material strength, the action combination should adopt the basic combination and it should meet the requirements of Equation 5-15:

$$N \leqslant A_p f_c \psi_c \qquad (5\text{-}15)$$

where

f_c ——the design value of axial compressive strength of concrete, kN/m²;

N ——design value of vertical force on pile top under basic combination, kN;

A_p ——cross-sectional area of pile body, m^2；

ψ_c ——coefficient of working conditions: 0.85 for precast piles, 0.9 for non-compacted piles in dry operation, 0.7-0.8 for bored piles with slurry wall protection and casing wall protection, and 0.6 for compacted piles in soft soil areas.

Example 5-1:

The plane layout of the pile and the size of the pile cap in a pile foundation is given in Figure 5-14. Under the standard combination, the vertical force $F=6000$ kN and the moment $M_x=M_y=1000$ kN · m is transmitted from the upper structure to the column. The size of the reinforced concrete precast pile is 400 mm×400 mm. The buried depth of the pile cap is 3.0 m, and the cross-sectional size of the column is 700 mm×700 mm, located in the center of the pile cap. The weighted average weight of the soil above the pile cap is taken as 20 kN/m^3。Please determine the minimum bearing capacity of a foundation pile that meets the bearing capacity requirements.

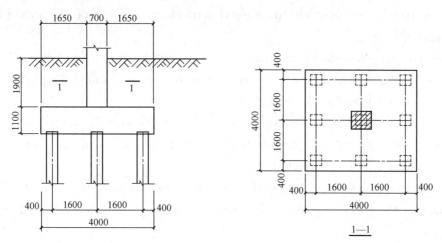

Figure 5-14　Pile arrangement in a pile foundation(mm)

Solution:

Calculate the average load applied on piles,

$$N_k = \frac{F_k+G_k}{n} = \frac{6000+4\times4\times3\times20}{9} = 773 \text{ kN}$$

Calculate the maximum load applied on a pile,

$$N_{kmax} = \frac{F_k+G_k}{n} + \frac{M_{xk}y_1}{\sum y_i^2} + \frac{M_{yk}x_1}{\sum x_i^2} = 773 + \frac{1000\times1.6}{1.6^2\times6} + \frac{1000\times1.6}{1.6^2\times6} = 981 \text{ kN}$$

Bearing capacity verification,

$$R_a \geqslant N_k = 773 \text{ kN}$$

$$R_a \geqslant \frac{N_{kmax}}{1.2} = \frac{981}{1.2} = 817 \text{ kN}$$

Adopt the greatervalue, the minimum bearing capacity of a foundation pile should be 817 kN.

5.6 Negative friction resistance of piles

5.6.1 Concept of negative friction of piles

Under the pressure of the pile top, the pile moves downward relative to the surrounding soil, so the soil exerts upward friction on the pile, which constitutes a part of the bearing capacity of the bearing pile. This kind of friction is usually called positive friction, or simply side friction or side resistance.

However, due to some reasons, the downward displacement of the pile itself is less than that of the surrounding soil, so that the friction acting on the pile is downward, which actually becomes a pull-down load acting on the pile side, which is called negative friction. Negative friction reduces the bearing capacity of the compressed pile, increases the load on the pile, and may lead to excessive settlement, so it should be checked when it is unavoidable.

There are many reasons for negative friction, some of which are as follows.

(1) A large area of large load is distributed on the ground around the pile, such as a large area of stacking load in the warehouse.

(2) The pile penetrates under-consolidated soft clay or new fill layer, the pile end is supported on the hard soil layer and the soil around the pile consolidates and settles with time under the action of self-weight.

(3) Due to the large-scale decline of groundwater (such as pumping a large amount of groundwater), the effective stress of the compressible soil layer increases and compression occurs.

(4) The self-weight collapsible loess sinks in water, the melting of frozen soil.

(5) Piling in sensitive soil causes structural damage to the soil around the pile, which leads to remodeling and consolidation. The figure shows the moment M of vane shear apparatus when the soil is damaged. A field test shows that the negative friction caused by this is about 17% of the undrained strength of clay.

5.6.2 Distribution of negative friction

The distribution range of negative friction on the pile body depends on the relative displacement between the pile body and the soil around the pile. Generally, except for non-long piles supported on bedrock, negative friction is not completely distributed along the pile body.

The ab line in Figure 5-15(b) represents the distribution of soil settlement around the pile with depth, s_e represents the settlement of surface soil. The cd line is the downward displacement curve of the pile body, which is calculated by $s_D = s_p + s_{se}$, where s_p is the settlement of the pile end, meaning the whole downward movement of the pile; s_{se} is the compression amount of pile material below this section. It can be seen that the intersection of the ab line and the cd line is at Point O. At Point O the displacement of the pile and the soil around the pile is equal, and there is no relative displacement and friction between them, so the Point O is called neutral point. Above the neutral point, the settlement of soil at each section is greater than the downward displacement of each point of the pile, so it is a negative friction zone; Below the neutral point, the settlement of the soil is less than the downward displacement of each point of the pile, so it is a positive friction zone. The neutral point is the dividing point of positive and negative friction, so it is the maximum point of pile axial force, which is shown in Figure 5-15(d). The distribution of frictional resistance acting on the pile side is shown in Figure 5-15(c).

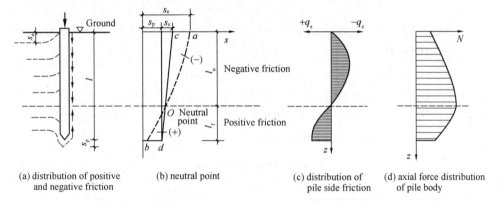

Figure 5-15　Negative friction distribution and neutral point of pile

The neutral point depth l_n is related to the compressibility and deformation conditions of soil around the pile, the distribution of soil layer and the stiffness of the pile. But in fact, it is difficult to determine the position of neutral point accurately. Obviously, the smaller the settlement at the pile tip s_p, the greater the l_n. When $s_p=0, l_n=l$, that is, the whole pile distributed negative friction. For piles with negative friction, the ratio of neutral point depth to pile length given in *Technical Code for Building Pile Foundations* (*JGJ 94—2008*) is shown in Table 5-4.

Neutral point depth l_n　　　　　　　　　　　　　　　Table 5-4

Property of bearing stratum	Cohesive soil and silt	Medium-dense sand	Gravel, pebble	Bedrock
Neutral Point depth ratio l_n/l_0	0.5-0.6	0.7-0.8	0.9	1.0

Note: 1. l_n and l_0 are the depth of neutral point from the top of pile and the lower limit depth of soft soil around pile.
2. When the pile passes through the self-weight collapsible loess layer, l_n will increase by 10% according to the listed value (except when the bearing layer is bedrock).
3. When the consolidation of soil around the pile and the consolidation of pile foundation are completed at the same time, take $l_n=0$.
4. When the calculated settlement of the soil around the pile is less than 20 mm, l_n shall be multiplied by the tabulated value 0.4-0.8.

The above neutral point depth l_n refers to the situation when the settlement of the pile and the surrounding soil is stable. Because the consolidation of the soil around the pile develops with time, the neutral point depth also changes with time.

5.6.3 Calculation of negative friction

It is difficult to accurately calculate negative friction of pile body due to many factors. Most scholars believe that the friction on the pile side is related to the effective vertical stress on the pile side. According to a large number of tests and engineering measurements, the "effective stress method" proposed by L. Bjerrum is closer to reality. It is used to calculate the standard value of negative friction:

$$q_{si}^n = K\tan\varphi'\sigma' = \xi_n\sigma' \tag{5-16}$$

where

K ——lateral pressure coefficient of soil, which can be taken as static earth pressure coefficient;

φ' ——effective stress internal friction angle of soil;

σ' ——vertical effective stress in the soil around the pile, kPa;

ξ_n ——the negative friction coefficient of the soil around the pile is related to the type and state of the soil. Please refer to Table 5-5.

Negative friction coefficient ξ_n Table 5-5

Soil type	ξ_n	Soil type	ξ_n
Saturated soft soil	0.15-0.25	Sand	0.35-0.50
Cohesive soil and silt	0.25-0.40	self-weight collapsing loess	0.20-0.35

Note: 1. In the same kind of soil, the larger value in the table is taken for the compaction pile, and the smaller value in the table is taken for the non-compaction pile.

2. According to its composition, the filled soil takes the larger value of similar soil in the table.

The total negative friction (pull-down load) Q_g^n on the pile side considering the pile group effect can be calculated by the following equations.

$$Q_g^n = \eta_n \cdot u \sum_{i=1}^{n} q_{si}^n l_i \tag{5-17}$$

$$\eta_n = s_{ax} \cdot s_{ay} \Big/ \left[\pi d\left(\frac{q_s^n}{\gamma_m} + \frac{d}{4}\right)\right] \tag{5-18}$$

where

n ——number of soil layers above neutral point;

l_i ——the thickness of the ith soil layer above the neutral point, m;

η_n ——negative friction coefficient of pile group effect;

s_{ax}, s_{ay} ——center distance of longitudinal and transverse piles respectively, m;

q_s^n ——standard value of weighted average negative friction of soil layer thickness around pile above neutral point, kPa;

γ_m ——weighted average weight of soil layer thickness around pile above neutral point (floating weight is taken below groundwater level), kN/m³.

For a single pile foundation or pile group effect coefficient calculated by the above formula $\eta_n > 1$, take $\eta_n = 1$.

The existence of negative friction reduces the bearing capacity of the pile and increases the load on the pile. Some measures can be taken to avoid or reduce negative friction in the design and construction of pile foundation. For precast reinforced concrete piles and steel piles, a layer of asphalt sliding layer with considerable viscosity can be coated on the pile body. For cast-in-place piles, plastic film is laid on the hole wall or high-consistency bentonite slurry is used to form a sliding layer on the pile wall before pouring concrete. Adjusting the construction order and taking measures to reduce the relative displacement of pile and soil after the building is used are also effective methods to reduce negative friction.

5.6.4 Bearing capacity verification of negative friction piles

In the design of pile foundation, the influence of negative friction on the bearing capacity and settlement of pile foundation can be considered according to the specific situation. When there is no engineering experience to refer to, the *Technical Code for Building Pile Foundations* (*JGJ 94—2008*) suggests verification according to the following equations.

(1) For friction piles, the upper resistance above the neutral point can be calculated as zero, and the bearing capacity of the pile can be verified according to the following equation:

$$N_k \leqslant R_a \qquad (5\text{-}19)$$

(2) For end-bearing piles, in addition to meeting the requirements of the above formula, the pull-down load Q_g^n caused by negative friction should be considered, and the bearing capacity of foundation piles can be verified by the following equation:

$$N_k + Q_g^n \leqslant R_a \qquad (5\text{-}20)$$

In the above two equations, only the side resistance and end resistance below the neutral point are used to calculate R_a.

(3) When the soil layer is uneven or the building is sensitive to uneven settlement, the pull-down load caused by negative friction should be included in the additional load to verify the settlement of pile foundation.

Example 5-2:

One pile is placed under one column. As Figure 5-16 shown, the pile has a diameter of $d = 850$ mm and length of $L = 22$ mm. The negative friction is caused due to large area of surcharge. Given the neutral point depth of $l_n/l_0 = 0.8$, the negative friction coefficient of silty clay $\xi_n = 0.2$, and negative friction coefficient of pile group effect $\eta_n = 1.0$, please de-

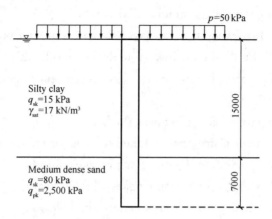

Figure 5-16 The pile and soil layer around the pile

termine the pull-down load Q_g^n on the pile.

Solution:

the neutral point depth of $l_n/l_0 = 0.8$, $l_n = 0.8 \, l_0 = 0.8 \times 15 = 12.0$ m

$$q_{si}^n = \xi_n \sigma_i' = 0.2 \times \left(\frac{1}{2} \times 7 \times 12 + 50\right) = 18.4 \text{ kPa}$$

$$q_{si}^n > q_{sk} = 15 \text{ kPa take } q_{si}^n = 15 \text{ kPa}$$

$$Q_g^n = \eta_n u \sum_{i=1}^{n} q_{si}^n l_i = 1 \times 3.14 \times 0.85 \times 15 \times 12 = 480.42 \text{ kN}$$

5.7 Uplift bearing capacity verification of piles

5.7.1 Uplift bearing capacity of piles

After the anti-floating piles of deep-buried light structures and underground structures, piles frozen and pulled out in frozen soil areas, and high-rise buildings are subjected to large overturning force, some or all of the piles often bear uplift force. So the uplift checking calculation of pile foundation should be carried out.

Different from the pressure-bearing pile, when the pile is subjected to pull-out load, the pile moves upward relative to the soil, which makes the stress state, stress path and deformation of the soil around the pile different from that of the pressure-bearing pile, so the pull-out friction is generally less than the compressive friction. Especially, the friction resistance in sand is much smaller than that in compression. However, in saturated clay, faster uplift can generate larger negative excess pore water pressure in the soil, which may make it more difficult to pull out the pile, but it will dissipate with time, so it is generally not included in the uplift force. There may be two pull-out situations for pile foundation under pull-out load, that is, each single pile is pulled out and the whole pile group (including soil between piles) is pulled out, depending on which situation provides less total resistance.

The most effective method to determine the uplift bearing capacity of a single pile is to carry out static pile uplift load test in the field for important buildings. For non-important buildings, when there is no local experience, the standard value of ultimate uplift bearing capacity T_{uk} of a single pile can be calculated according to Equation 5-21:

$$T_{uk} = \sum \lambda_i q_{sik} u_i l_i \tag{5-21}$$

when the pile group is pulled out as a whole, the ultimate uplift bearing capacity T_{gk} of each pile in the pile group can be calculated according to Equation 5-22:

$$T_{gk} = \frac{1}{n} u \sum \lambda_i q_{sik} l_i \tag{5-22}$$

where

λ_i ——uplift reduction coefficient of the ith layer of soil, please refer to Table 5-6 for values;

u_i ——perimeter of pile body, for piles with equal diameter, $u=\pi d$, for piles with enlarged bottom, in the range $l_i=(4\text{-}10)d$ above the pile bottom; $u=\pi D$, the greater the internal friction angle of soil, the higher the value of l_i, m;

u ——peripheral circumference of pile group, m;

q_{sik} ——the ultimate side resistance of the ith layer of soil when pile is compressed, kPa.

Pullout coefficient λ　　　　　　　　　　　　Table 5-6

Soil group	λ
Sand	0.5-0.7
Cohesive soil and silt	0.7-0.8

Note: When the ratio of pile length l to pile diameter d is less than 20, take the smaller value as λ.

5.7.2　Verification of uplift bearing capacity of piles

The uplift calculation of a single pile can be carried out by Equation 5-23:

$$N_k \leqslant \frac{T_{uk}}{2} + G_p \tag{5-23}$$

where

N_k ——uplift force on single pile corresponding to the effect of standard combination, kN;

G_p ——the standard value of the dead weight of a single pile, and the buoyancy is deducted below the groundwater level, kN.

At this time, the uplift calculation of single pile can be carried out by the following equation:

$$N_k \leqslant \frac{T_{gk}}{2} + G_{gp} \tag{5-24}$$

where

N_k ——pullout force of single pile calculated according to combined effect of standard action;

G_{gp} ——the total dead weight of the pile-soil in the volume surrounded by the pile group foundation is divided by the total number of piles n, and the buoyancy is deducted from the groundwater level, kN.

5.8 Lateral bearing capacity verification of piles

5.8.1 Behavior of piles under lateral loads

For industrial and civil construction projects, most pile foundations mainly bear vertical compressive loads, but sometimes they also bear lateral loads, such as instantaneous wind load, crane braking load and earthquake load. Since the lateral load in most cases is not large, for the convenience of construction, vertical piles are often used to resist lateral forces also. Therefore, the situation and bearing capacity of vertical piles under lateral loads are discussed below.

The displacement of the pile under the lateral load will cause the soil around the pile to deform and produce resistance. When the lateral load is low, this resistance is mainly provided by the soil near the ground, and the deformation of the soil is mainly elastic compression deformation. With the increase of the load, the deformation of the pile will also increase, and the surface soil will gradually yield plastically, so that the lateral load will be transferred to the deeper soil layer.

When the deformation increases to the extent that the pile cannot allow, or the soil around the pile loses stability, the lateral ultimate bearing capacity of the pile is reached.

The lateral bearing capacity of a single pile, like the vertical compressive bearing capacity, should meet the following three requirements.

(1) The soil around the pile will not lose stability.

(2) The pile body will not break.

(3) The normal use of the building will not be affected by the excessive lateral displacement of the pile top.

Obviously, whether it can meet the requirements depends on the soil conditions around the pile, the depth of the pile, the section stiffness of the pile, the material strength of the pile and the nature of the building. The better the soil quality, the deeper the pile penetrates into the soil, the greater the resistance of the soil and the higher the lateral bearing capacity of the pile. Piles with poor flexural performance, such as cast-in-place piles with low reinforcement ratio, are often damaged by the fracture of the pile body, while piles with good flexural performance, such as reinforced concrete piles and steel piles, are often controlled by the properties of the surrounding soil. In order to ensure the normal use of buildings, according to engineering experience, the lateral displacement of pile top should be controlled not to be greater than 10 mm, while the buildings sensitive to lateral displacement should not be greater than 6 mm.

Other factors that affect the lateral bearing capacity of a single pile are the embedding condition of the pile top and the interaction of the piles in the pile group. When there is a rigid cap constraint, the pile top cannot rotate, but can only translate. Under the same lateral load, it reduces the lateral displacement of the cap and increases the bending moment of the pile top. Figure 5-17 shows the schematic diagram of deformation and failure behavior of different types of piles under the condition of whether the pile top is embedded or not. The influence of pile groups is as follows: under the constraint of a rigid cap, lateral load causes lateral displacement of each pile, and the gap left by the displacement of the front row of piles reduces the resistance of the rear row of piles. This effect is particularly significant when the number of piles is large and the pile spacing is small.

In addition, when the soil around the pile cap is undisturbed or the backfill is compacted, the lateral resistance of the surrounding soil to the pile cap can be included, which reduces the lateral load acting on the pile.

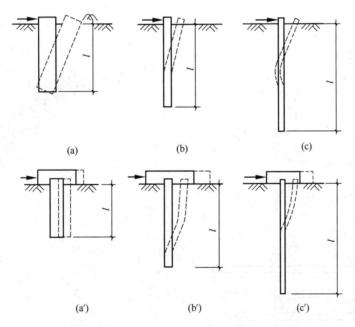

(a) and (a') rigid piles; (b) and (b') semi-rigid piles; (c) and (c') flexible piles
(a), (b) and (c) free pile top; (a'), (b') and (c') embedded pile top
Figure 5-17 Schematic diagram of failures of various piles

5.8.2 Static lateral load tests of a single pile

For important buildings with large lateral load, the characteristic value of lateral bearing capacity of a single pile should be determined by static lateral load test of a single pile. Figure 5-18 is the schematic diagram of the test. First of all, two identical test piles can be made on site, and a jack for loading is placed laterally between the two piles. Cyclic loading and unloading method is often used in the loading method, so as to be consistent with the instantaneous and repeated lateral load carried by the pile foundation. For pile foundation subjected to long-term lateral load, slow loading method can also be used for test.

Before the test, $\frac{1}{15}$-$\frac{1}{10}$ of the predicted lateral ultimate bearing capacity of a single pile is taken as the loading increment of each stage. After each level of load is applied, keep the constant load for 4 min to measure the lateral displacement, then unload to zero, and stop for 2 min to measure the residual lateral displacement, do this five times so as to complete a loading and unloading cycle of the first level of load. Then the next level of load test is carried out, repeat this test for 10-15 levels of load. The test shall not stop halfway until the pile body is broken or the lateral displacement increases sharply under dead load, or when the lateral displacement exceeds 30-40 mm, then test may be terminated.

According to the lateral load test, draw the lateral load-time-displacement (H_0-t-x_0) curve, and take the previous load whose curve drops sharply as the ultimate load H_u(kN), as shown in Figure 5-19.

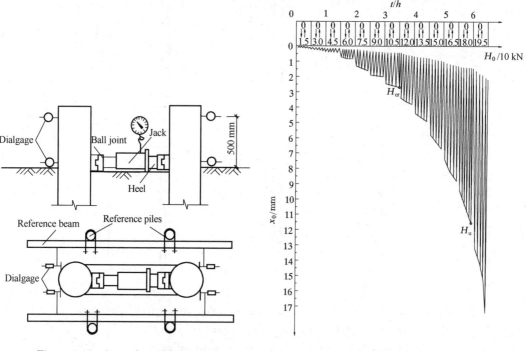

Figure 5-18 Static lateral load test device for a single pile

Figure 5-19 Lateral load test curve (H_0-t-x_0 curve)

When cracks are allowed in the pile, the ultimate lateral bearing capacity divided by the safety factor of 2.0 can be taken as the characteristic value R_{ha} of the lateral bearing capacity of a single pile. For precast reinforced concrete piles, steel piles and cast-in-place piles with full-section reinforcement ratio of not less than 0.65%, 75% of the load corresponding to the lateral displacement of 10 mm on the ground (6 mm for buildings sensitive to lateral displacement) can also be taken as the characteristic value of lateral bearing capacity of single pile according to the static load test results; When cracks are not allowed in the pile, the lateral critical load is taken 0.75 time of the statistical value is the characteristic value of single grain lateral bearing capacity. Take the primary load before the pile body breaks as the ultimate load. If there is no obvious abrupt change and broken pile, the terminal load of the second straight line section of the lateral force-displacement gradient ($H_0 - \Delta X_0/\Delta H_0$) curve can be taken as the ultimate load, as shown in Figure 5-20.

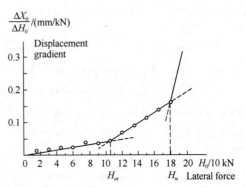

Figure 5-20 $H_0 - \Delta X_0/\Delta H_0$ curve

5.8.3 Theoretical analysis of elastic long pile under lateral loads

The lateral deformation and bearing capacity of elastic long piles are usually analyzed by the method of lateral elastic foundation beam. The Winkler subgrade model is used to study the deflection curve of particles under the combined action of lateral load and soil resistance at the side of pile, which is shown in Figure 5-21(a). Through the differential equation of deflection curve, the bending moment, shear force and deformation of each section of pile can be obtained. The differential equation of the deflection curve of pile is

$$EI \frac{d^4 x}{dz^4} = -p_x \tag{5-25}$$

where

p_x ——lateral resistance of soil acting on pile, kN/m; As hypothesized by Winkler subgrade model, it can be calculated as follow:

$$p_x = k_h x b_0 \tag{5-26}$$

where

b_0 ——calculated width of pile, m, as shown in Table 5-7;

x ——lateral displacement of pile, m;

k_h ——lateral resistance coefficient of soil, or called lateral subgrade coefficient, kN/m³.

Calculation width of pile section b_0 Table 5-7

Section width b or diameter d/m	Round pile	Square pile
>1	$0.9(d+1)$	$b+1$
≤1	$0.9(1.5d+0.5)$	$1.5b+0.5$

The magnitude and distribution of lateral resistance coefficient k_h directly affects the solution of the above differential Equation 5-25, the internal force of cross section and the calculation of pile deformation. The k_h is related to the type of soil and the depth of pile penetration. Because of the different hypothesis about the distribution of k_h, there are different calculation and analysis methods, and the four assumptions shown in Figure 5-21 are adopted. The general expression is as follows:

$$k_h = mz^n \tag{5-27}$$

(1) The constant method. It assumes that k_h along the pile depth is constant (Figure 5-21b), that is $n=0$ as shown in Equation 5-27.

(2) The k method. It assumes that k_h above the first zero point z_t of the deflection surface curve increases in a straight line ($n=1$) or parabola ($n=2$) along the depth, otherwise, it is constant ($n=0$), as shown in Figure 5-21(c).

(3) The m method. It assumes k_h exhibits a proportional increase with depth (Figure 5-21d), that is $n=1.0$ as shown in Equation 5-27.

(4) c method. It assumes k_h exhibits a parabolic variation with depth (Figure 5-21e), that is $n=0.5$, $m=c_0$, in Equation 5-27.

The measured data show that when the lateral displacement of a pile is large, the calculation results of the m method are close to reality in most cases, thus m method is widely used in China. The m value in Equation 5-27 varies with soil type and soil state.

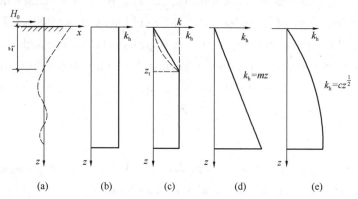

(a)　　　(b)　　　(c)　　　(d)　　　(e)

Figure 5-21　Deformation and lateral resistance coefficient of piles

5.8.4　Lateral bearing capacity verification

When the external force acting on the pile foundation is mainly lateral force, the lateral bearing capacity of the pile foundation should be checked. In a pile foundation composed of piles with the same cross section, it can be assumed that the lateral force H_{ik} on each pile is the same, that is,

$$H_{ik} = \frac{H_k}{n} \tag{5-28}$$

where

H_k——lateral force acting on the bottom of the pile cap under the action standard combination, kN;

H_{ik} ——the lateral force acting on the ith single pile, kN.

Under the action of lateral load, the bearing capacity of the pile should be verified using the following equation:

$$H_{ik} \leqslant R_{ha} \tag{5-29}$$

where

R_{ha} ——characteristic value of lateral bearing capacity of a single pile, kN.

5.9 Settlement of pile foundations

Although the settlement of pile foundation compared with shallow foundation on natural foundation is lower. However, with the increase of the scale and size of buildings and the improvement of settlement requirements, the settlement calculation of pile foundation is also needed in many cases. Like the settlement calculation of shallow foundation, the final settlement calculation of pile foundation should use quasi-permanent load combination. The basic calculation method is still the layered summation method based on the assumption of unidirectional compression, homogeneous isotropy and elasticity of soil.

At present, there are two main types of layered summation methods for pile settlement calculation. One is the equivalent deep foundation method, and the other is the Mindlin stress calculation method. These two methods are introduced as follows.

5.9.1 Equivalent deep foundation method

The essence of the equivalent deep foundation method is to take the pile end plane as the surface of elastic body and calculate the additional stress of each point below the pile end with Boussinesq solution. The settlement is then calculated with the same one-way compression layered summation method as that of shallow foundation. The equivalent deep foundation means that a certain range of pile caps, piles and soil around piles are regarded as an equivalent solid deep foundation. The compression deformation from the ground to the pile end plane is not considered. This kind of method is suitable for pile spacing $s \leqslant 6d$.

There are two hypotheses about how to apply the upper additional load to the pile tip plane. One is that the load spreads along the outside of the pile group as shown in Figure 5-22(a), and the other is to deduct the side friction around the pile group as shown in Figure 5-22(b). The first method will be described as follows.

As Figure 5-22(a) shown, the diffusion angle is taken as $\dfrac{1}{4}$ of the weighted average friction angles in each soil layer through which the pile passes. The additional pressure p_0 at the pile end plane can be calculated by Equation 5-30:

$$p_0 = \frac{F + G_T}{\left(b_0 + 2l \times \tan\frac{\overline{\varphi}}{4}\right)\left(a_0 + 2l \times \tan\frac{\overline{\varphi}}{4}\right)} - p_c$$

(5-30)

where

F —— vertical force acting on the top surface of pile cap when quasi-permanent combination is applied, kN;

G_T —— in the area after diffusion, the total weight of pile caps, piles and soil from the pile end plane to the design ground, which can be calculated according to 20 kN/m³, kN;

a_0, b_0 —— long and short sides of rectangular area of outer edge of pile group, m;

$\overline{\varphi}$ —— weighted average value of internal friction angle of soil layer through which the pile passes, (°);

l —— buried depth of pile, m;

p_c —— the dead weight pressure ($l + d$ depth) of foundation soil on the pile end plane, kPa, and the buoyancy should be deducted for the part below the groundwater level.

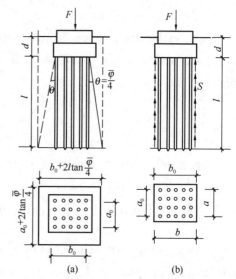

Figure 5-22   Bottom area of solid deep foundation

Sometimes, the difference between the total weight of pile-soil mixture along pile length l and the total weight of in-situ foundation soil with the same volume can be ignored, and the approximate calculation can be made by Equation 5-31:

$$p_0 = \frac{F + G - p_{c0} \times a \times b}{\left(b_0 + 2l \times \tan\frac{\overline{\varphi}}{4}\right)\left(a_0 + 2l \times \tan\frac{\overline{\varphi}}{4}\right)} \quad (5\text{-}31)$$

where

G —— the self-weight of the pile cap and the soil on the pile cap, which can be calculated according to 20 kN/m³, and the buoyancy is deducted from the underwater part, kN;

p_{c0} —— the dead weight pressure of foundation soil at the elevation of the bottom surface of the bearing platform, and the buoyancy is deducted from the part below the groundwater level, kPa;

a, b —— length and width of pile cap, m.

After calculating the additional pressure p_0 at the pile end plane, the settlement can be calculated by layered summation method according to the area after diffusion:

$$s = \psi_p \sum_{i=1}^{n} \frac{p_i h_i}{E_{si}} \quad (5\text{-}32)$$

where

s —— final calculated settlement of pile foundation, mm;

n —— calculate the number of layers;

E_{si} —— compressive modulus of the ith layer of soil from deadweight stress to deadweight stress plus additional stress, MPa;

h_i —— thickness of the ith layer under the plane of pile end, m;

p_i —— average vertical additional stress of the ith layered soil under the plane of pile tip, kPa;

ψ_p —— the empirical coefficient of pile foundation settlement calculation, which can be determined according to the statistical data of local projects in different regions. It can also refer to Table 5-8, in which $\overline{E}_s$ can be referred to in Equation 3-18.

Empirical coefficient of pile foundation settlement calculation ψ_p　　　Table 5-8

$\overline{E}_s$/MPa	≤15	25	35	≥45
ψ_p	0.5	0.4	0.35	0.25

5.9.2　Mindlin-Geddes Method

According to the load transfer characteristics of piles, the total load Q acting on the top of a single pile is decomposed into two parts: pile end resistance $Q_p(=\alpha Q)$ and pile side resistance $Q_s[=(1-\alpha)Q]$. The side resistance of pile Q_s can be divided into the evenly distributed frictional resistance $Q_{s1}(=\beta Q)$ and the linearly increased frictional resistance $Q_{s2}[=(1-\alpha-\beta)Q]$ as shown in the Figure 5-23, where α and β are the end resistance and evenly distributed frictional resistance that accounts for the proportion of the total load respectively. Correspondingly, according to Mindlin's solution, Gades used Q_p, Q_{s1} and Q_{s2} to obtain the formula for calculating the additional stress generated in the foundation soil. These formulas can be used to calculate the additional stress caused by various piles in the foundation, and then calculate the settlement of pile foundation. This method is called Mindlin-Gades method, or simply Mindlin method.

The coefficients of α and β shall be determined statistically according to the measured data of local projects. For general friction piles, it can be assumed that the pile side resistance increases linearly along the pile body, that is $\beta=0$. In this way, the vertical additional stress of each friction pile at a certain point in the foundation is the sum of the vertical additional stresses σ_{zp} and σ_{zs} generated by the pile end load Q_p and the pile side load Q_s. If there are m piles, the additional stress generated by each pile at this point is superimposed one by one, and calculated by the following equation:

$$p_i = \sum_{k=1}^{m}(\sigma_{zp,k} + \sigma_{zs,k}) \qquad (5-33)$$

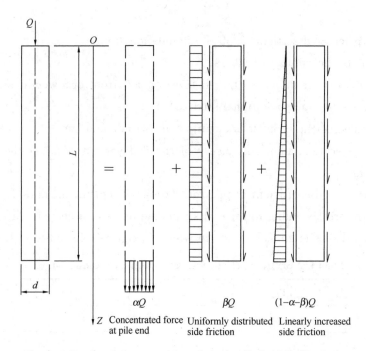

Figure 5-23 Load decomposition of single pile for Mindlin-Gades

where

p_i ——additional stress generated at the midpoint of the ith soil layer, kPa;

$\sigma_{zp,k}$ ——additional stress at the midpoint of the ith soil layer generated by the kth pile end load, which can be calculated using Equation 5-34, kPa.

$$\sigma_{zp,k} = \frac{\alpha Q}{l^2} I_{p,k} \qquad (5-34)$$

For general friction piles, it can be assumed that the frictional resistance of the pile is all distributed in a triangle type along the pile body, that is $\beta=0$. The additional stress generated by the pile side load of the kth pile at this point is:

$$\sigma_{zs,k} = \frac{Q}{l^2}(1-\alpha) I_{s2,k} \qquad (5-35)$$

where

l ——the length of the pile in the soil, m;

I_p, I_{s2} ——stress influence coefficient, which can be derived by integrating Mindlin stress equation, in which:

I_p is the stress influence coefficient of the concentrated force at the pile bottom, which is derived by integration.

$$I_p = \frac{1}{8\pi(1-\nu)} \left\{ \frac{(1-2\nu)(m-1)}{A^3} - \frac{(1-2\nu)(m-1)}{B^3} + \frac{3(m-1)^3}{A^5} \right. $$
$$\left. + \frac{3(3-4\nu)m(m+1)^2 - 3(m+1)(5m-1)}{B^5} + \frac{30m(m+1)^3}{B^7} \right\} \qquad (5-36)$$

I_{s2} is the stress influence coefficient when the distributed load on the pile side increases linearly along the pile body, which is deduced by integration as follows:

$$I_{s2} = \frac{1}{4\pi(1-\nu)}\left\{\frac{2(2-\nu)}{A} - \frac{2(2-\nu)(4m+1) - 2(1-2\nu)(1+m)m^2/n^2}{B}\right.$$
$$- \frac{2(1-2\nu)m^3/n^2 - 8(2-\nu)m}{F} - \frac{mn^2 + (m-1)^3}{A^3}$$
$$- \frac{4m^2 m + 4m^3 - 15n^2 m - 2(5+2\nu)(m/n)^2(m+1)^3 + (m+1)^3}{B^3} \quad (5\text{-}37)$$
$$- \frac{2(7-2\nu)mn^2 - 6m^3 + 2(5+2\nu)(m/n)^2 m^3}{F^3} - \frac{6mn^2(n^2 - m^2) + 12(m/n)^2(m+1)^5}{B^5}$$
$$\left. + \frac{12(m/n)^2 m^5 + 6mn^2(n^2 - m^2)}{F^5} + 2(2-\nu)\ln\left(\frac{A+m-1}{F+m} \times \frac{B+m+1}{F+m}\right)\right\}$$

$A^2 = n^2 + (m-1)^2$; $B^2 = n^2 + (m+1)^2$; $F^2 = n^2 + m^2$; $n = r/l$; $m = z/l$

where

ν —— Poisson's ratio of foundation soil;

r —— calculate the lateral distance between the point and the axis of the pile;

z —— calculate the vertical distance between the stress point and the bottom of the pile cap.

Substituting $\sigma_{zp,k}$ and $\sigma_{zs,k}$ into Equation 5-33, the additional stress caused by m piles at this point can be obtained. Then, the settlement is still calculated according to the layered summation method of unidirectional compression. That is, the settlement is calculated by substituting p_i into the Equation 5-32, which can be expressed as

$$s = \psi_{pm} \frac{Q}{l^2} \sum_{i=1}^{n} \frac{\Delta h_i}{E_{si}} \sum_{k=1}^{m} [\alpha I_{p,k} + (1-\alpha) I_{s2,k}] \quad (5\text{-}38)$$

At this time, the empirical coefficient ψ_{pm} of pile foundation settlement calculation should be statistically determined according to the measured data of local projects. It can also be obtained using Table 5-9.

Empirical coefficient ψ_{pm}　　　　　　　　　　　　　Table 5-9

$\overline{E}_s$/MPa	≤15	25	35	≥40
ψ_{pm}	1	0.8	0.6	0.3

The Mindlin stress formula is more consistant with the actual situation of pile load transfer than the Boussinesq solution, but the calculation formula is slightly complicated.

5.10　Structure design of pile caps

5.10.1　Basic requirements for the structure of pile caps

The pile cap can be divided into independent pile cap under column or wall, strip pile cap beam under column or wall, raft cap and box cap of pile-raft foundation and pile-box foundation.

The size of pile cap is related to the number and spacing of piles, and should be determined through comprehensive economic and technical comparison. Its size mainly meets the requirements of bending, punching and shearing resistance. According to the basic requirements, the minimum thickness and the minimum width of the pile caps are not less than 300 mm and 500 mm respectively. The distance between the edge of the bearing platform and the center of the side pile is not less than the diameter of the pile or the side length of the pile; the distance between the outer edge of the pile and the edge of the cap is not less than 150 mm, and the distance between the outer edge of the pile and the edge of the cap beam is not less than 75 mm for the strip cap beam.

In order to ensure the integrity of the connection between the pile group and the pile cap, the length of the pile top embedded in the pile cap should not be less than 50 mm, and it should not be less than 100 mm for large diameter piles. The anchorage length of the main reinforcement inserted into the pile cap shall not be less than 35 times the diameter of the main reinforcement. For large-diameter cast-in-place piles, when one column and one pile are used, pile caps can be set or piles and columns can be directly connected.

The concrete strength grade of the bearing platform shall not be lower than C20. The thickness of concrete protective layer of longitudinal reinforcement is not less than 70 mm. When placing concrete cushion, the protective layer shall be not less than 50 mm, and shall not be less than the length of the pile head embedded in the pile cap. The reinforcement of rectangular pile cap is arranged in two-way uniform length. The steel bars of the three-pile cap are uniformly arranged according to the three-way strip, and the triangle surrounded by the innermost three steel bars should be within the cross section of the pile, as shown in Figure 5-24.

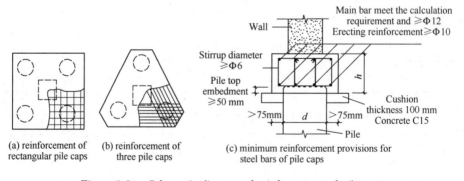

Figure 5-24 Schematic diagram of reinforcement of pile cap

In addition to meeting the requirements of design and calculation, the general reinforcement still needs to meet the minimum reinforcement ratio specified in the *Code for Design of Concrete Structures* (GB 50010—2010). The diameter of the main reinforcement is not less than 12 mm, the vertical reinforcement is not less than 10 mm and the

5.10 Structure design of pile caps

stirrup is not less than 6 mm, as shown in Figure 5-24(c).

A single pile in a column with seismic requirements and two independent pile caps are often connected by connecting beams. The buried depth of the pile cap is generally not determined by the bearing capacity of the soil layer at the bottom of the pile cap, but mainly considers the structural design and environmental conditions of the building, and can be buried as shallow as possible when these conditions are met. For the pile foundation with seismic requirements, in order to increase the lateral resistance, the buried depth of the pile cap can be increased. The pile cap is located on a good soil layer, which can play the joint role of pile cap, pile and soil, increase the bearing capacity of pile group foundation and reduce settlement. Backfill around the pile caps should be compacted by layers of plain soil, lime soil and graded sand, or concrete pile caps should be poured in the original pit.

The bending, punching and shearing resistance of the pile caps should be calculated. When the concrete strength grade of the pile caps is lower than that of the column or pile, the local compressive bearing capacity of the pile caps under the column or on the pile should be checked.

5.10.2 Bending calculation of pile caps

If the thickness of the pile cap is small and the amount of reinforcement is insufficient, the pile cap may be bent under the force transmitted by the column. The test and engineering practice show that the pile cap of independent pile foundation under the column is beam failure. The flexural cracks appear in two directions parallel to the column edge, and the maximum bending moment occurs at the column edge, as shown in Figure 5-25.

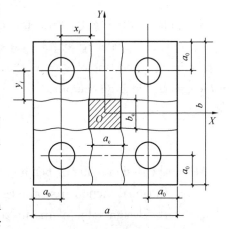

Figure 5-25 Bending moment failure mode of four-pile cap

1. Multi-pile rectangular cap

For multi-pile rectangular cap, the calculation section is taken at the column edge and the change of cap section. As Figure 5-26 shown, the bending moment can be calculated as Equations 5-39 and 5-40.

$$M_x = \Sigma N_i y_i \tag{5-39}$$

$$M_y = \Sigma N_i x_i \tag{5-40}$$

where

M_x, M_y —— design value of bending moment at the calculated section around the X axis and Y axis direction respectively, kN · m;

x_i, y_i —— the distance from the center point of the ith pile perpendicular to the Y axis and the X axis direction to the corresponding calculated section, m;

N_i —— the design value of the vertical net reaction force of the ith pile under the basic combination, without considering the self-weight of the pile cap and its above fill soil, kN.

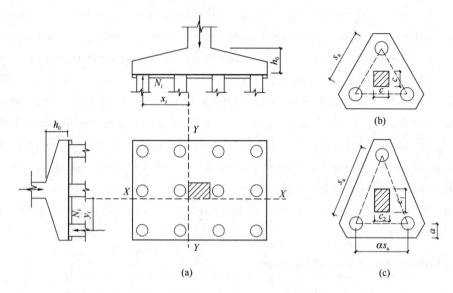

Figure 5-26 Schematic diagram of moment calculation of pile caps

2. Equilateral triangular three-pile cap

For three-pile caps, triangular caps are usually used, which can be divided into equilateral triangles and isosceles triangles. The design bending moment should be calculated by the greatest load on the pile.

$$M = \frac{N_{max}}{3}\left(s_a - \frac{\sqrt{3}}{4}c\right) \tag{5-41}$$

where

M —— the design value of bending moment of plate and strip within the range of orthogonal section from the centroid of bearing platform to each edge, kN · m;

N_{max} —— the design value of the maximum vertical force of single pile among the three piles under the basic combination of action, excluding the pile cap and its soil weight, kN;

s_a —— pile spacing, m;

c —— the side length of a square column, when it is a cylinder, take $c = 0.8d$ (d is the diameter of the cylinder), m.

3. Isosceles triangle three-pile cap

$$M_1 = \frac{N_{max}}{3}\left(s_a - \frac{0.75}{\sqrt{4-\alpha^2}}c_1\right) \tag{5-42}$$

$$M_2 = \frac{N_{max}}{3}\left(\alpha s_a - \frac{0.75}{\sqrt{4-\alpha^2}}c_2\right) \tag{5-43}$$

where

M_1, M_2——design value of bending moment of plate and strip within the orthogonal section range from the centroid of the bearing plat form to the two waist edges and bottom edges of the bearing platform, kN·m;

s_a——long pile spacing, m;

α——the ratio of short pile spacing to long pile spacing, when $\alpha < 0.5$ it can be designed according to the two-pile cap with variable cross section;

c_1, c_2——side length of column section perpendicular to and parallel to the bottom of pile cap, m.

5.10.3 Punching calculation of independent pile caps under columns

The thickness of pile cap is often determined by punching calculation. There are two main types of punching failure of pile caps. One is from the edge of columns and at the step change position of the pile cap, and the punching failure cone occurs when the tension crack along the inclined plane of pile cap is $\geqslant 45°$. The other is the top of the corner pile forms an upward semi-cone punching failure with angle $\geqslant 45°$ relative to the edge of the cap, as shown in Figure 5-27.

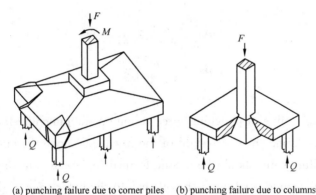

(a) punching failure due to corner piles (b) punching failure due to columns

Figure 5-27 Punching failure of pile cap

1. Punching of pile cap due to column

There are two possible forms of punching failure of the column to the pilecap, that is, punching failure along the column edge or along the step change of the pile cap. The most dangerous punching cone is the case where the included angle between the cone and the bottom of the cap is $\geqslant 45°$. The inclination angles of this cone in different directions may be different, as shown in Figure 5-28. At this time, the punching calculation can be carried out according to the following equation:

$$F_l \leqslant 2[\beta_{0x}(b_c + a_{0y}) + \beta_{0y}(a_c + a_{0x})]\beta_{hp} f_t h_0 \qquad (5\text{-}44a)$$

$$F_l = F - \Sigma N_i \qquad (5\text{-}44b)$$

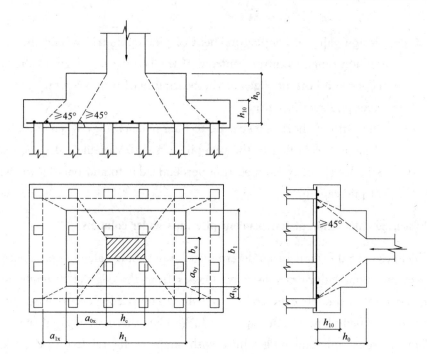

Figure 5-28 Punching calculation of column to pile cap

$$\beta_{0x} = \frac{0.84}{\lambda_{0x} + 0.2} \tag{5-44c}$$

$$\beta_{0y} = \frac{0.84}{\lambda_{0y} + 0.2} \tag{5-44d}$$

where

F_l——the design value of punching force acting on the punching failure cone, without considering the self-weight of the filled soil above the pile cap. The punching failure cone should be a cone formed by connecting the column edge or the step change position of the pile cap to the inner edge of the corresponding pile top, and the angle between the cone and the bottom surface of the pile cap is $\geqslant 45°$, kN;

f_t——design value of tensile strength of concrete of pile cap, kN/m^2;

h_0——the effective height of the punching failure cone is generally the thickness of the punching bearing capacity section of the bearing platform minus the thickness of the protective layer, m;

β_{hp}——the section height influence coefficient of punching shear bearing capacity, which is the same as β_{hp} in Equation 3-25;

β_{0x}, β_{0y}——punching coefficient;

λ_{0x}, λ_{0y}——the ratio of puncture to span, $\lambda_{0x} = \frac{a_{0x}}{h_0}$, $\lambda_{0y} = \frac{a_{0y}}{h_0}$, which shall meet the requirements of 0.25-1.0;

a_{0x}, a_{0y}——the lateral distance from the column edge or step change in x, y direction to the inner edge of the corresponding pile, m;

F——design value of axial force at column under basic load combination, kN;

$\sum N_i$——the sum of the design values of vertical net reaction force of each pile within the range of punching failure cone bottom surface, kN.

2. Punching calculation of corner pile to pile cap

Because the same pile type is assumed to share the total vertical force under the cap according to the linear law, a corner pile will bear the maximum net vertical force under eccentric load. On the other hand, when the corner pile is punched upwards, the cone of punching resistance is only half, that is, for four prism cap, there are only two punching surfaces. Undoubtedly, punching of the corner piles is often the most dangerous.

(1) Punching calculation of corner piles for multi-pile rectangular caps

This situation is shown in Figure 5-29. In Figure 5-29(a), the pile cap is tapered; In Figure 5-29(b), the pile cap is stepped. For the case of Figure 5-29(a), the cone height of the punching inverted cone is related to the punching cone angle. On one hand, the calculation height is complicated, and on the other hand, the impact section Δh_0 of the extra part is not very reliable, so h_0 is still taken as the effective height of the outer edge of the

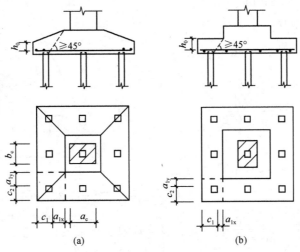

Figure 5-29 Corner pile punching to pile cap

pile cap, which is safe. The corresponding calculation equation is as follows:

$$N_l \leqslant \left[\beta_{1x}\left(c_2 + \frac{a_{1y}}{2}\right) + \beta_{1y}\left(c_1 + \frac{a_{1x}}{2}\right) \right] \beta_{hp} f_t h_0 \qquad (5\text{-}45a)$$

$$\beta_{1x} = \frac{0.56}{\lambda_{1x} + 0.2} \qquad (5\text{-}45b)$$

$$\beta_{1y} = \frac{0.56}{\lambda_{1y} + 0.2} \qquad (5\text{-}45c)$$

where

N_l——the design value of vertical force of the corner pile corresponding to the basic combination of applied load, without considering the self-weight of the pile cap and its fill, kN;

β_{1x}, β_{1y}——punching coefficient of corner pile;

$\lambda_{1x}, \lambda_{1y}$——angle-particle impact-span ratio, whose value satisfies 0.25-1.0, $\lambda_{1x} = \dfrac{a_{1x}}{h_0}$, $\lambda_{1y} = \dfrac{a_{1y}}{h_0}$;

c_1, c_2——the distance from the inner edge of the corner pile to the outer edge of the pile cap, m;

a_{1x}, a_{1y}——the lateral distance from the intersection point of the 45° punching line and the top surface of the pile cap from the inner edge of the corner pile at the bottom of the pile cap, or the intersection point at the step change of the pile cap to the inner edge of the corner pile, m;

h_0——effective height of the outer edge of the bearing platform, m.

(2) Punching calculation of corner piles with three triangular caps

The specific calculation is as shown in Figure 5-30, and is calculated as follows:

Bottom corner pile,

$$N_l \leqslant \beta_{11}(2c_1 + a_{11})\tan\dfrac{\theta_1}{2}\beta_{hp} f_t h_0 \quad (5\text{-}46a)$$

$$\beta_{11} = \dfrac{0.56}{\lambda_{11} + 0.2} \quad (5\text{-}46b)$$

Top corner pile,

$$N_l \leqslant \beta_{12}(2c_2 + a_{12})\tan\dfrac{\theta_2}{2}\beta_{hp} f_t h_0 \quad (5\text{-}47a)$$

$$\beta_{12} = \dfrac{0.56}{\lambda_{12} + 0.2} \quad (5\text{-}47b)$$

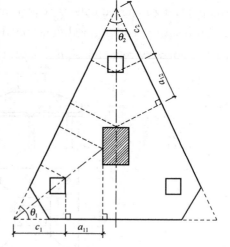

Figure 5-30 Punching calculation of corner piles of triangular cap

where

$\lambda_{11}, \lambda_{12}$——angular particle impact-span ratio, $\lambda_{11} = \dfrac{a_{11}}{h_0}$, $\lambda_{12} = \dfrac{a_{12}}{h_0}$, shall meet the requirements 0.25-1.0;

a_{11}, a_{12}——The lateral distance from the 45° intersection of the punching tangent line and the top surface of the pile cap to the inner edge of the corner pile from the inner edge of the corner pile at the bottom of the pile cap to the adjacent pile cap edge. When the column edge is within the 45° line, take the connecting line between the column edge and the inner edge of the pile as the cone line of the punching cone, m.

In the above punching calculation, the cylindrical and circular piles can be converted into square columns and square piles, and the conversion formula is $b=0.8d$.

5.10.4 Shear calculation of independent pile cap of pile foundation

For the independent pile cap under the column, the shear bearing capacity of the inclined section of the pile cap should be checked. The shear plane is an inclined section formed by connecting the column (wall) edge with the inner edge of the pile, as shown in Figure 5-31. At this time, the shear calculation should be carried out for the inclined section formed by the connecting line between the column (wall) edge and the pile edge, the step change and the pile edge respectively. When there are multiple rows of piles on the column (wall) to form multiple inclined sections, each inclined section should also be checked. The calculation equation is as follows:

$$V \leqslant \beta_{hs} \alpha f_t b h_0 \tag{5-48a}$$

$$\alpha = \frac{1.75}{\lambda + 1.0} \tag{5-48b}$$

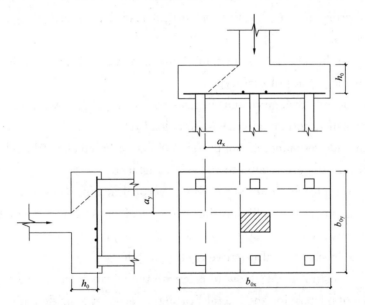

Figure 5-31 Shear calculation of inclined section of pile cap

where

V——the maximum shear design value of the inclined section under the basic combination of action, without taking into account the self-weight of the pile cap and its fill, which is equal to the sum of the corresponding vertical net reaction forces of the piles outside the inclined section, kN;

b——the calculated width at the calculated section of the bearing platform, the calculated width at the step change position of the bidirectional stepped bearing platform and the calculated width of the bidirectional tapered bearing platform shall be converted, m;

h_0 ——effective height of the cap at the calculated width, m;

α ——shear coefficient;

β_{hs} ——the influence coefficient of shear bearing capacity section height, calculated according to $\beta_{hs} = \left(\dfrac{800}{h_0}\right)^{\frac{1}{4}}$, in which, when h_0 is less than 800 mm, it is taken as 800 mm, and when h_0 is greater than 2000 mm, it is taken as 2000 mm;

λ ——calculate the shear span ratio of the section, $\lambda_x = \dfrac{a_x}{h_0}, \lambda_y = \dfrac{a_y}{h_0}, a_x, a_y$ are the lateral distances from the column edge or the step change of the pile cap to the calculated pile edge of a row of piles. When $\lambda < 0.25$; take $\lambda = 0.25$, when $\lambda > 3.0$, take $\lambda = 3.0$.

Questions

5-1   Please describe the classification of piles according to their use function.

5-2   Please describe the classification of vertical loaded piles according to their bearing characteristics.

5-3   According to the pile-forming method, how many types of piles can be divided? What are the characteristics of each type?

5-4   What are the side resistance and tip resistance of a pile? How do these two kinds of resistance come into play after the pile is loaded?

5-5   How is the side resistance of compression pile distributed along the pile body in sandy soil? What factors are related to the side resistance?

5-6   How to determine the bearing capacity of vertical bearing piles?

5-7   How to determine the uplift bearing capacity of piles? For piles of the same size, which bearing capacity is larger by comparing compression pile and uplift pile?

5-8   What is the mechanism of negative friction of piles?

5-9   What is the neutral point? How to calculate the negative friction of piles?

5-10 What factors determine the lateral bearing capacity of a single pile?

5-11 How to determine the lateral bearing capacity of a single pile?

5-12 What are the calculation methods of pile foundation settlement? What are the main differences?

5-13 What are pile group effect and pile group effect coefficient?

5-14 One pile cap under column has a buried depth of 2.5 m, a bottom size of 4 m×4 m, and C30 concrete. The column section size is 1.0 m×1.0 m. Four underwater bored piles with a diameter of $d = 800$ mm are used, and the layout is shown in Figure 5-32. Corresponding to the standard load combination, $F_k = 6,068$ kN and $M_k = 408$ kN·m, (permanent load control). The average internal friction of the soil layer is $\bar{\varphi} = 12°$. Please finish the following questions.

① Calculate the characteristic value of the vertical bearing capacity of a single pile.

② Verify the bearing capacity of a single pile in pile foundation.

③ Calculate the maximum bending moment of the pile cap.

④ Verify the punching failure of the column.

⑤ Verify the punching of corner piles.

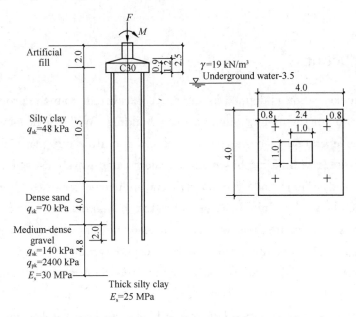

Figure 5-32 The layout of piles and soil layers (unit: m)

Chapter 6   Retaining and Protection Structures for Foundation Excavation

6.1   Introduction

Foundation excavation is defined as the space excavated from the ground downwards for the construction of the underground part of a building. With the development of urban construction, underground space has been developed and utilized in major cities around the world, such as high-rise building basements, underground warehouses, underground civil defense works, and various underground civil and industrial facilities. For example, the World Trade Center in the United States is located in Manhattan District, the center of New York. Its excavation area reaches 65000 m^2 and the total excavation earthwork is about 12 million m^3. At that time, there were two underground subway lines under construction. Therefore, the designers finally adopted the diaphragm wall and the anchored retaining structure deep into the rock stratum.

In China, the construction of subways and high-rise buildings has resulted in a large number of deep excavation projects. The foundation excavation of Shanghai Jinmao Tower is a deep foundation excavation in soft soil foundation. Its excavation area reaches 20000 m^2. The excavation depth of its main building is 19.65 m and the diaphragm wall depth is 36 m with the wall thickness of 1.0 m.

In the total cost of high-rise buildings in China, the subgrade and foundation parts often account for 1/4-1/3. They cost more when foundation excavation is constructed in complex geological regions. The construction period of foundation excavation often accounts for more than 1/3 of the total construction period. The foundation excavation is the key to ensure the successful completion of the foundation engineering of the main building. On one hand, it should ensure the stability of the soil and retaining structure of the foundation excavation itself. On the other hand, it must ensure the safety and normal use of surrounding buildings, underground facilities, pipelines and roads. It should also be noted that because foundation excavation is generally a temporary project, there is often a lot of room for saving cost and shortening construction period in design and construction. Therefore, foundation excavation is not only risky, but also highly flexible and creative.

In China, the practice of high-rise buildings and underground engineering is developing rapidly, but the corresponding theory and technology lag behind the engineering practice. This can be seen from two aspects: on one hand, the design is conservative, which results in a waste of financial resources and time. On the other hand, frequent accidents in foundation excavation have caused great economic losses and casualties. Figure 6-1 shows the various foundation excavation failure cases in China.

Figure 6-1   Foundation excavation failure cases in China

The theoretical difficulties affecting the accurate design of foundation excavation mainly include the following aspects.

(1) Earth pressure calculation on retaining and protection structures

A large number of field monitoring data in different regions show that the internal force in the retaining structure calculated according to the traditional earth pressure theory is often greater than the measured value. This is mainly because the magnitude and distribution of the earth pressure acting on the pre-set retaining structure are influenced by many factors such as the properties of the undisturbed soil, the deformation of the retaining structure, the three-dimensional effect of the foundation excavation, the stress state and stress path of the soil, and etc. They are quite different from the earth pressure acting on the retaining wall by the artificial fill behind the wall. At present, the earth pressure on retaining and protection structures is still difficult to accurately analyze it.

(2) The occurrence form and movement of water in soil

With the increase of excavation depth, it may involve several layers of groundwater with different occurrence forms, such as upper stagnant water, phreatic water and confined water. Excavation, drainage and dewatering of foundation excavation will cause complicated groundwater seepage, which not only increases the difficulty of calculating the water pressure and earth pressure on the retaining structure, but also makes the seepage stability of foundation excavation become a problem that must be solved in deep foundation excavation.

(3) Influences of foundation excavation on the surrounding environment

If the stability analysis of limit equilibrium is used to ensure the safety of foundation excavation itself, it is often necessary to carry out deformation evaluation when analyzing the influence of foundation excavation on adjacent buildings, underground facilities and pipelines. It should be noted that the difficulty of deformation prediction is much higher than that of stability analysis.

To solve these three difficulties, it relies on the development of geotechnical mechanics theory and the engineering practice. In recent years, many new technologies and design methods have been developed in the deep foundation excavation practice of high-rise buildings in China.

6.2 Foundation excavation and supporting methods

6.2.1 Safety grade of foundation excavation and retaining structures

The foundation excavation and retaining structures should meet the following functional requirements.

(1) Ensure the safety and normal use of buildings (structures), underground pipelines, and roads around the foundation excavation.

(2) Ensure the construction space of the main underground structure.

When designingfoundation excavation and retaining structures, various factors such as the complexity of the surrounding environment, geological conditions, as well as the excavation depth should be considered. The safety grade of retaining structures should be adopted according to Table 6-1. For the same foundation excavation, different safety grades can be used for different parts.

Safety grade of retaining structures   Table 6-1

Safety grade	Consequence of failure
A	Failure of retaining structure and excessive deformation of the surrounding soil have very serious impacts on the surrounding environment or the construction safety of main structures
B	Failure of retaining structure and excessive deformation of the surrounding soil have serious impacts on the surrounding environment or the construction safety of main structures
C	Failure of retaining structure and excessive deformation of the surrounding soil have no serious impacts on the surrounding environment or the construction safety of main structures

6.2.2 Types of foundation excavation and retaining structures

Whether to adopt retaining structure in foundation excavation and what kind of retaining structure should be determined are significantly depended on the surrounding environment, the main building and underground structure, excavation depth, engineering geology and hydrogeology, construction equipment and construction season, and etc. Table 6-2 shows the selection of main excavation and retaining structures.

Application of various retaining structures Table 6-2

Structure type		Applicable conditions		
		Safety grade	Depth of foundation excavation, environmental conditions, soil types and groundwater conditions	
Retaining structure	Anchored retaining structure	A B C	Deep foundation excavation	1. The soldier pile wall is suitable for the foundation excavation with dewatering or cutoff curtain. 2. The diaphragm wall should be used as the external wall of the main underground structure, and can also be used for intercepting water. 3. Anchor should not be used in soft soil layer, gravel soil and sandy soil layer with high water level. 4. When there are buildings, basements, underground structures, etc. adjacent to the foundation excavation, the effective bolt length of the anchor is insufficient. The anchor should not be used. 5. When the anchor construction will cause damage to the buildings around the foundation excavation or violate the urban underground space planning and other regulations, the anchor should not be used
	Strutted retaining structure		Deep foundation excavation	
	Cantilever retaining structure		Shallow foundation excavation	
	Double-row-piles wall		When anchored, strutted and cantilever structures are not applicable, double-row-piles wall can be considered.	
	Top-down method for combining support structure with main structure		Deep foundation excavation with complicated surrounding conditions.	
Soil nailing wall	Single soil nailing wall	B C	Non-soft soil foundation excavation above groundwater level or after precipitation, and the excavation depth should not be greater than 12 m	When there are buildings and important underground pipelines in the potential sliding surface of foundation excavation, soil nailing wall is not suitable
	Prestressed anchor composite soil nailing wall		Non-soft soil foundation excavation above groundwater level or after precipitation, and the excavation depth should not be greater than 15 m	

Chapter 6 Retaining and Protection Structures for Foundation Excavation

continued

Structure type		Applicable conditions	
		Safety grade	Depth of foundation excavation, environmental conditions, soil types and groundwater conditions
Soil nailing wall	Cement-soil pile composite soil nailing wall	B C	When used in non-soft soil foundation excavation, the excavation depth of foundation should not be greater than 12 m. When used in muddy soil foundation excavation, the excavation depth should not be greater than 6 m. It should not be used in gravel soil and sandy soil with high water level
	Micro-pile composite soil nailing wall		It is suitable for the foundation excavation above the groundwater level or after precipitation. When it is used for non-soft soil foundation excavation, the excavation depth should not be greater than 12 m. When used in muddy soil foundation excavation, the excavation depth should not be greater than 6 m
Gravity cement-soil wall		B C	It is suitable for muddy soil and silt foundation excavation, and the excavation depth should not be greater than 7 m
Slope excavation		C	1. The construction site should meet the conditions of slope setting. 2. It can be combined with the aboveretaining structure

Note (rightmost column for Soil nailing wall rows): When there are buildings and important underground pipelines in the potential sliding surface of foundation excavation, soil nailing wall is not suitable

The various types of foundation excavation and retaining structures will be presented as follows.

1. Retaining structure

The retaining structure mainly composed of retaining components (support piles, diaphragm wall) and anchor or supports, or only retaining components. The retaining structure can be divided into four types as Figure 6-2 shown.

(1) Cantilever retaining structure. No internal support or anchor are installed, which makes construction inside the foundation excavation convenient. Due to the small stiffness of the support piles, both internal force and deformation are relatively large. When the environmental requirements are high, it is not suitable to excavate deep foundation pits. The excavation depth should not exceed 5 m in soft soil sites.

(2) Strutted retaining structure. Internally supported pile wall retaining structure. Setting up single or multiple internal supports can effectively reduce the internal force and deformation of the retaining wall. By setting up multiple supports, it can be used to excavate deep foundation pits. However, setting up internal supports brings significant inconvenience to the excavation of earthwork and the construction of underground structures.

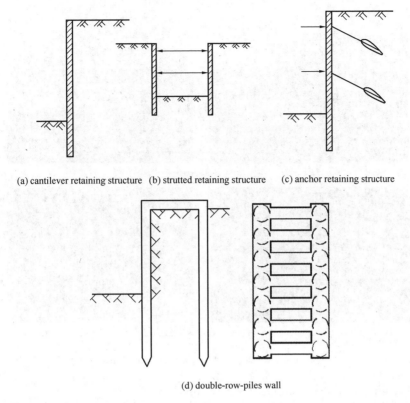

(a) cantilever retaining structure (b) strutted retaining structure (c) anchor retaining structure

(d) double-row-piles wall

Figure 6-2　Types of retaining structures

The internal support can be horizontal or inclined.

(3) Anchor retaining structure. By fixing single-layer or multi-layer soil anchor rods in stable soil layers to reduce the internal force and deformation of the retaining wall. Multi-layer anchor rods can be used to excavate deep foundation pits.

(4) Double-row-piles wall. The double-row-piles wall consists of two rows of supporting piles and beams connected by these two rows of piles.

The support piles can be steel sheet piles, bored cast-in-place piles, and etc. The various kinds of support piles are shown in Figure 6-3. The steel sheet pile is a simple retaining wall, which is composed of overlapping or side by side buckles on the front and back sides of the steel. It has the advantages of good durability and secondary utilization. However, it cannot prevent water and small particles in the soil. Measures such as water isolation or precipitation should be taken in areas with high groundwater levels. Meanwhile, it usually has significant deformation after excavation due to its weak bending resistance and low stiffness. The bored cast-in-place pile has the advantage of high strength, high stiffness, good support stability and small deformation. However, the gaps between piles can easily cause soil erosion, especially in areas with soft clay at high water levels. Construction measures such as grouting, cement mixing piles, and rotary jet grouting piles can be taken to solve the water prevention problem.

(a) steel sheet piles

(b) bored cast-in-place piles

Figure 6-3 Various kinds of support piles

2. Soil nailing wall and composite soil nailing wall

(1) Soil nailing wall

Soil nailing wall is composed of soil nail and shotcrete surface, which is shown in Figure 6-4. Among them, soil nail is the main stress component, which is made by inserting a slender metal bar (usually steel bar) into the soil wall and drilling (digging) in advance. Soil nailing wall is suitable for general cohesive soil, silty soil, miscellaneous fill and plain fill, non-loose sandy soil, gravel soil and etc. It is not suitable for pebble and gravel layer with larger particle size, because it is difficult to drill (dig) holes in this soil layer. It is also not suitable for saturated soft clay sites. Dewatering measures should be taken when the excavation bottom is below the groundwater level.

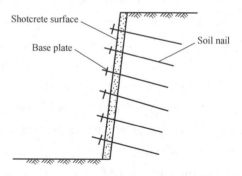

Figure 6-4 Soil nailing wall support

(2) Composite soil nailing wall

Soil nailing wall is widely used due to its low cost, convenient earthwork excavation and short construction period. However, it cannot be used in soft clay, the soil layer below groundwater without artificial precipitation and sites with strict requirements on the subsidence. In engineering practice, the soil nailing wall is often used by combining with other engineering techniques, which is called the composite soil nailing wall. At present, soil nailing wall is mainly combined with prestressed anchor, micro-pile and cement-soil pile or cement-soil curtain. Figure 6-5 is a schematic diagram of two different forms of composite soil nailing walls.

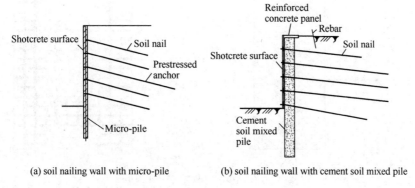

(a) soil nailing wall with micro-pile   (b) soil nailing wall with cement soil mixed pile

Figure 6-5   Composite soil nailing wall

(3) Shotcrete and anchor support

On the surface, no obvious difference can be seen between shotcrete and anchor support and soil nailing wall. However, their reinforcement mechanisms are quite different. The structure of anchor is shown in Figure 6-6. The anchor is the main component to bear stress. Each anchor is divided into anchoring section and free section. And the anchoring section is located outside the sliding surface of soil and the pressure grouting is adopted. The free section is within the sliding surface of soil and the free section is not grouted. Generally, the bold rod of the anchor is made of steel strand or finish rolled threaded steel bar. Generally, the bold rod is tensioned by prestress.

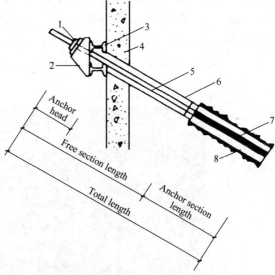

Figure 6-6   Structure of anchor

1 – anchor; 2 – pedestal; 3 – waist beam; 4 – support pile wall;
5 – anti – corrosion mortar; 6 – drilling; 7 – anchor bar;
8 – anchor solid

3. Gravity cement-soil wall

The gravity cement-soil wall is a grid or continuous wall formed by several rows of overlapping cement-soil piles constructed on the outside of the excavation by deep mixing method or high-pressure jet grouting method, as Figure 6-7 shown. The gravity cement-soil wall uses its self-weight to resist the earth pressure behind the wall and it is often treated as a gravity retaining wall. The gravity cement-soil wall is suitable for soft soil subgrade. It should not be used in fill layers with a large amount of crushed stones, bricks, and other organic matter. The permeability coefficient of cement-soil is relatively small. Therefore, the gravity cement-soil wall has good waterproof performance and does not require additional waterproof curtains.

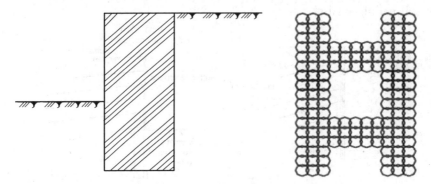

Figure 6-7 Gravity cement-soil wall

4. Diaphragm wall support

The diaphragm wall support is formed by using special trenching equipment to dig a slotted hole, then puts a reinforcing cage in the slotted hole and pours concrete. It has high rigidity and good integrity, and the deformation of the surrounding soil caused by excavation is small. Figure 6-8 shows the diaphragm wall. It can be used in deep excavation, however its cost is high.

Figure 6-8 The diaphragm wall support

5. Top-down method

The top-down method is a top-down construction method with the beams, slabs and columns of the underground structure of the main project as the support for excavation. In this method, the retaining structure and the permanent basement structure are integrated, and the temporary retaining structure is not required. The construction speed can be accelerated, and the underground and upper structures can be constructed at the same time.

However, the excavation face is narrow and the excavation is limited, so the joints of columns, walls, beams and slabs need to be properly handled. Figure 6-9 is a schematic diagram of the top-down construction of a subway station under the road.

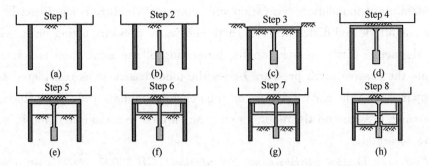

Step 1: construct retaining structure; Step 2: construct middle column of the main structure; Step 3: construct roof; Step 4: backfill and recover the road; Step 5: excavate the soil; Step 6: construct upper main structure; Step 7: excavate the lower soil; Step 8: construct lower main structure

Figure 6-9 A schematic diagram of the top-down method of a subway station

6. Slope excavation

The excavation without any retaining structure is slope excavation. When conditions permit, slope excavation is the most economical and quick excavation method of foundation excavation. The following conditions need to be met when adopting this excavation method: 1) The soil should be general cohesive soil or silty soil, dense macadam soil or weathered rock. 2) The groundwater level is low or artificial precipitation measures are adopted. 3) The site has space for sloping and for stacking soil materials, machines and tools and traffic roads around the foundation excavation. Meanwhile, the slope excavation will not have adverse effects on adjacent buildings and municipal facilities.

The *Code for Design of Building Foundation* (*GB 50007—2011*) specified the allowable soil slope ratio which is given in Table 6-3. Slope excavation can be used alone or often combined with other retaining structures.

Allowable soil slope ratio Table 6-3

Soil Type	State	Allowable slope ratio (ratio of height to width)	
		Slope height within 5 m	Slope height between 5-10 m
Gravel soil	Dense	1 : 0.50-1 : 0.35	1 : 0.75-1 : 0.50
	Median-dense	1 : 0.75-1 : 0.50	1 : 1.00-1 : 0.75
	Slight-dense	1 : 1.00-1 : 0.75	1 : 1.25-1 : 1.00
Cohesive soil	Hard	1 : 1.00-1 : 0.75	1 : 1.25-1 : 1.00
	Hard-plastic	1 : 1.25-1 : 1.00	1 : 1.50-1 : 1.25

Note: 1. The filling material in gravel soil should be hard or hard-plastic clayey soil.
2. For sandy soil or gravel soil with sandy filling, the allowable slope ratio should be determined according to the natural repose angle.

6.3 Calculation of earth and water pressure on retaining structures

The stability of foundation excavation and retaining structure is significantly dependent on the magnitude and distribution of earth and water pressure acting on it. Generally speaking, the active earth pressure on the outer side of the excavation is considered as loads, while the passive earth pressure below the inner basement is considered as resistance. Compared with the earth pressure on the general retaining wall, the influencing factors of the earth pressure on the retaining structure are more complicated and it is difficult to calculate accurately.

The *Code for Design of Building Foundation* (GB 50007—2011) stipulates that when checking the stability of retaining structures, the earth pressure can generally be calculated according to active earth pressure or passive earth pressure using Coulomb or Rankine earth pressure theory. However, when there are strict restrictions on the horizontal displacement of the retaining structure, the static earth pressure should be used for calculation. When the retaining structure is designed according to the deformation control, the earth pressure acting on the retaining structure should be calculated according to the interaction principle between the retaining structure and the soil.

1. Calculation of earth pressure above groundwater level

In the stability calculation of the retaining structure, the active earth pressure outside the excavation and the passive earth pressure inside the pit are generally calculated according to Rankine earth pressure theory, as shown in Figure 6-10.

The active earth pressure p_{aj} at the depth z_j below the ground behind the retaining structure is

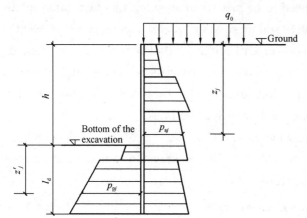

Figure 6-10 Schematic diagram of earth pressure calculation on retaining structure

$$p_{aj} = K_{aj}\left(q_0 + \sum_{i=1}^{j} \gamma_i h_i\right) - 2c_j \sqrt{K_{aj}} \qquad (6\text{-}1)$$

$$K_{aj} = \tan^2\left(45° - \frac{\varphi_j}{2}\right) \qquad (6\text{-}2)$$

where

φ_j —— internal friction angle of at the depth z_j, °;

c_j —— cohesion of soil at the depth z_j, kPa;

q_0 —— uniformly distributed load on the ground behind the wall, kPa;
γ_i —— the unit weight of the ith layer of soil, kN/m³;
h_i —— thickness of the ith layer of soil, m.

The passive earth pressure p_{pj} at the depth z'_j under the bottom of the excavation:

$$p_{pj} = K_{pj} \sum_{i=1}^{j} \gamma_i h_i + 2 c_j \sqrt{K_{pj}} \tag{6-3}$$

$$K_{pj} = \tan^2 \left(45° + \frac{\varphi_j}{2} \right) \tag{6-4}$$

2. Calculation of water pressure and earth pressure below groundwater level

According to the effective stress principle, the vertical stress of a point can be expressed as

$$\sigma_z = \sigma'_z + u \tag{6-5}$$

In this way, the active earth pressure at this point is

$$p_a = K_a \sigma'_z - 2c \sqrt{K_a} \tag{6-6}$$

water pressure is

$$p_w = u \tag{6-7}$$

6.4 Stability evaluation of foundation excavation

Most of the foundation excavation accidents are produced due to the instability of the excavation. The stability evaluation of foundation excavation belongs to the limit state design of bearing capacity. The instability of the excavation can be resulted from the insufficient shear strength of soil or retaining structure or the seepage failure. It should be noted that water in soil is often the main factor that causes the instability of foundation excavation.

6.4.1 Stability of pile or wall retaining structure

1. Anti-overturning stability evaluation

The embedding depth l_d below the bottom of the excavation is significantly depended on its anti-overturning stability. Figure 6-11 shows anti-overturning stability for two types of the retaining structure: (a) cantilever retaining structure and (b) anchored or strutted retaining structure. For the cantilever retaining structure, its embedded depth should meet the anti-overturning stability of the whole pile and wall relative to the bottom point O. The anchored or strutted retaining structure should meet the anti-overturning stability relative to the lowest anchor or strut point O. The anti-overturning stability evaluation is given in the Equation 6-8:

$$\frac{\sum M_{E_p}}{\sum M_{E_a}} \geq K_e \tag{6-8}$$

Chapter 6 Retaining and Protection Structures for Foundation Excavation

where

K_e —— safety factor of anti-overturning stability of pile and wall retaining structure. The K_e is 1.25, 1.2 and 1.15 corresponding to the safety grade of retaining structures A, B and C;

ΣM_{E_a} —— total overturning moment in active area, including overturning moment of active earth pressure and water pressure difference on both sides, kN · m;

ΣM_{E_p} —— total anti-overturning moment in passive zone, kN · m.

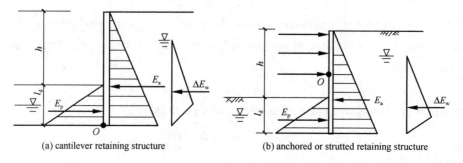

(a) cantilever retaining structure (b) anchored or strutted retaining structure

Figure 6-11 Anti-overturning stability of pile and wall retaining structure

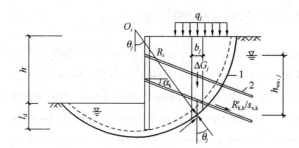

Figure 6-12 Overall stability of retaining structure

2. Overall stability evaluation

The overall stability of retaining structure can be calculated using Swedish circular Slice Method, as shown in Figure 6-12. When the circular sliding slice method is adopted, its overall stability shall meet the following requirements. For the cantilever retaining structure, the last item $R'_{k,k}$ in Equation 6-9 is zero. It is necessary to calculate the overall stability of various sliding surfaces with different centers and radii, and the minimum safety factor should meet the requirements.

$$\frac{\Sigma\{c_j l_j + [(q_j b_j + \Delta G_j)\cos\theta_j - u_j l_j]\tan\varphi_j\} + \Sigma R'_{k,k}[\cos(\theta_k + \alpha_k) + \psi_v]/s_{x,k}}{\Sigma(q_j b_j + \Delta G_j)\sin\theta_j} \geqslant K_s$$

$$\psi_v = 0.5\sin(\theta_k + \alpha_k)\tan\varphi \qquad (6-9)$$

where

K_s —— safety factor of overall stability of circular arc sliding;

c_j, φ_j —— cohesion and internal friction angle of soil at the sliding arc surface of the jth strip, kPa and °;

b_j —— the width of the jth strip, m;

θ_j —— the angle between the radius at the midpoint of the jth sliding arc surface and

the vertical line, °;
l_j —— length of sliding arc section of the jth strip, m, $l_j = b_j/\cos\theta_j$;
q_j —— the additional distributed load acting on the jth strip, kPa;
ΔG_j —— the self-weight of the jth strip, kN;
u_j —— pore water pressure on the sliding arc surface of the jth strip, kPa;
$R'_{k,k}$ —— the ultimate tensile force of the kth anchor rod on the circular sliding body, kN;
α_k —— the angle of the kth layer anchor rod, °;
θ_k —— the angle between the normal of the sliding surface at the kth anchor rod and the vertical line, °;
$s_{x,k}$ —— horizontal spacing of the kth anchor rod, m;
ψ_v —— calculation coefficient of anti-sliding moment caused by normal force of anchor rod on sliding surface;
φ —— internal friction angle of the kth anchor rod and sliding arc, °.

3. Pit bottom uplift stability evaluation

When the pit bottom is saturated soft clay, pit bottom uplift may occur if the embedding depth l_d is insufficient. Figure 6-13 shows the schematic view of the pit bottom uplift stability. The pit bottom uplift will cause series of problems such as the horizontal displacement of the retaining structure and the ground settlement behind the wall. The pit bottom uplift stability should meet the requirements of Equation 6-10.

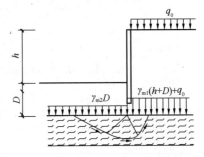

Figure 6-13 Pit bottom uplift stability evaluation

$$\frac{\gamma_{m2}DN_q + cN_c}{\gamma_{m1}(h+D) + q_0} \geq K_b$$

$$N_q = \tan^2\left(45° + \frac{\varphi}{2}\right)e^{\pi\tan\varphi} \qquad N_c = (N_q - 1)/\tan\varphi \qquad (6-10)$$

where

K_b —— safety factor of pit bottom uplift stability;
N_c —— bearing capacity coefficient, $N_c = 5.14$;
c —— undrained shear strength of soil below the wall bottom, kPa;
γ_{m1}, γ_{m2} —— unit weight of soil separated by the retaining structure, kN/m³;
D —— embedding depth of the retaining structure, m;
h —— depth of excavation, m;
q_0 —— ground load, kPa.

4. Seepage stability evaluation

When a suspended cutting off water curtain is used or there is a confined water aquifer

Chapter 6　Retaining and Protection Structures for Foundation Excavation

with water head higher than the pit bottom, the seepage stability evaluation should be carried out according to two different cases.

(1) As Figure 6-14a shown, when there is a confined water aquifer below the pit bottom with its water head higher than the pit bottom, and the hydraulic connection inside and outside the pit is not separated by a cutting off water curtain, the seepage stability can be conducted using Equation 6-11.

$$\frac{D\gamma}{h_w \gamma_w} \geqslant 1.1 \tag{6-11}$$

where

γ—— unit weight of soil from the top surface of the confined water aquifer to the bottom of the pit, kN/m^3;

h_w—— water head of the confined water aquifer, m.

(2) As Figures 6-14b and c shown, if the aquifer is homogeneous and the bottom of the suspended cutting off water curtain is located in a gravel soil, sandy soil, or silt aquifer, the seepage stability can be conducted using Equation 6-12.

$$\frac{(2l_d + 0.8D_1)\gamma'}{\Delta h \gamma_w} \geqslant K_f \tag{6-12}$$

where

K_f—— safety factor of seepage stability. The K_f is 1.6, 1.5 and 1.4 corresponding to

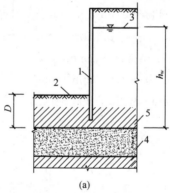

(a)

1- cutting off water curtain; 2- pit bottom; 3- water level of confined water aquifer;
4- confined water aquifer; 5- aquiclude

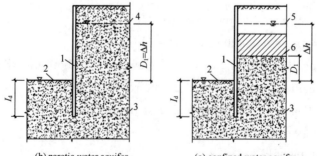

(b) paretic water aquifer　　　(c) confined water aquifer

1 - cutting off water curtain; 2 - pit bottom; 3 - aquifer; 4 - water level of paretic water aquifer;
5 - water level of confined water aquifer; 6 - top of confined water aquifer

Figure 6-14　Seepage stability evaluation

the safety grade of retaining structures A, B and C;

l_d —— insertion depth of cutting off curtain below the pit bottom, m;

D_1 —— thickness of soil layer from top surface of phreatic water or confined water aquifer to pit bottom, m;

γ' —— buoyant unit weight of soil, kN/m³;

Δh —— head difference inside and outside the foundation excavation, m;

γ_w —— the unit weight of water, kN/m³.

When the upper part is an impermeable layer and there is a confined water aquifer at a certain depth under the pit, the pit bottom n may also uplift, as shown in Figure 6-15. This phenomenon is called inrushing. Anti-inrushing stability can be conducted using Equation 6-13.

$$\frac{\gamma_m(t+\Delta t)}{p_w} \geqslant 1.1 \qquad (6\text{-}13)$$

where

γ_m —— natural unit weight of soil below the pit bottom and above the confined water aquifer, kN/m³;

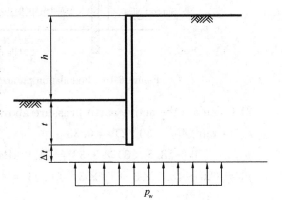

Figure 6-15　Anti-inrushing stability for confined water aquifer

$t+\Delta t$ —— depth from top surface of confined water aquifer to pit bottom, m;

p_w —— water pressure at the top of confined water aquifer, kPa.

Example 6-1:

As Figure 6-16 shown, the excavation depth of a foundation pit is 10 m, with a uniformly distributed load of $q_0=20$ kPa on the top of the slope. The groundwater outside the pit is located 6 m below the surface. The support piles and water-sealing curtain are used inside the pit. The support piles adopt bored piles with the diameter of 800 mm and length of 15 m. Please calculate the active lateral total pressure acting on the support piles.

Solution:

The distribution of the active pressure on support piles is shown in Figure 6-17. The detailed calculation is given as follows.

1) Calculate the active earth pressure in soil layer ①

$k_{a1} = \tan^2(45° - 20°/2) = 0.49$

$p_A = 20 \times 0.49 = 9.8$ kPa

$p_{B上} = (20 + 18.5 \times 3) \times 0.49 = 37$ kPa

$E_{a1} = \frac{1}{2} \times (9.8 + 37) \times 3 = 70.2$ kN/m

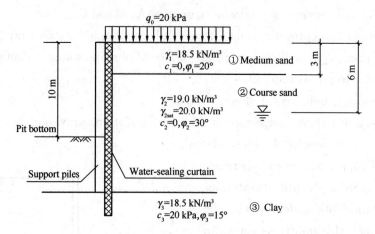

Figure 6-16 Foundation pit and distribution of soil layers

2) Calculate the active earth pressure above the ground water in soil layer ②

$k_{a2} = \tan^2(45° - 30°/2) = 0.33$

$p_{BF} = (20 + 18.5 \times 3) \times 0.33 = 24.9$ kPa

$p_c = (20 + 18.5 \times 3 + 19 \times 3) \times 0.33 = 43.7$ kPa

$E_{a2} = \dfrac{1}{2} \times 3 \times (24.9 + 43.7) = 102.9$ kN/m

3) Calculate the active earth pressure below the ground water in soil layer ②

$p_D = (20 + 18.5 \times 3 + 19 \times 3 + 10 \times 9) \times 0.33 = 73.4$ kPa

$E_{a3} = \dfrac{1}{2} \times 9 \times (43.7 + 73.4) = 527$ kN/m

The total active earth pressure on support piles:

$E_a = 70.2 + 102.9 + 527 = 700.1$ kN/m

Water pressure on support piles: $E_w = 9 \times 90/2 = 405$ kN/m

The total lateral pressure on support piles: $E = 700.1 + 405 = 1105.1$ kN/m

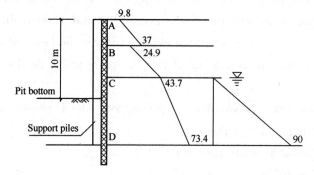

Figure 6-17 Distribution of the active pressure on support piles

6.4.2 Stability of gravity cement-soil wall

Gravity cement-soil wall is often used in soft cohesive soil. In addition to checking the arc sliding resistance, pit bottom uplift resistance and seepage stability introduced above, the following stability evaluation should be conducted as follows.

Figures 6-18(a) and (b) are a schematic diagram for the stability evaluation of gravity cement-soil wall. The width of gravity cement-soil wall is mainly determined by its overturning stability. In order to ensure sufficient stability against overturning, gravity cement-soil wall should have enough wall width b and embedding depth l_d. Generally, The l_d is $(0.8$-$1.2)h$ and the b is $(0.6$-$0.8)h$ with h of the wall height.

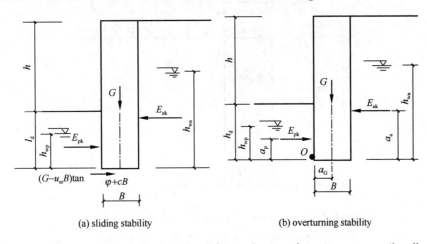

(a) sliding stability   (b) overturning stability

Figure 6-18 Schematic diagram of stability evaluation of gravity cement-soil wall

As Figure 6-18(a) shown, in order to ensure that the gravity cement-soil wall has enough horizontal sliding resistance along the wall bottom surface, it should meet the following requirement in Equation 6-14.

$$\frac{E_{pk}+(G-u_m B)\tan\varphi+cB}{E_{ak}} \geqslant 1.2 \qquad (6\text{-}14)$$

where

E_{pk}, E_{ak} —— total passive earth pressure and total active earth pressure, kN;

u_m —— water pressure acted in the wall bottom surface per unit width, kN/m;

G —— wall weight, kN;

B —— wall width, m;

φ —— friction angle of the soil at the bottom of the wall, °;

c —— cohesion of the soil at the bottom of the wall, kPa.

The overturning stability should be conducted according to Equation 6-15. The moment relative to the wall toe point O is calculated.

$$\frac{E_{pk}a_p+(G-u_m B)a_G}{E_{ak}a_a} \geqslant 1.3 \qquad (6\text{-}15)$$

where

a_p, a_a —— arm of passive earth pressure and active earth pressure to point O, m;

a_G —— arm of wall weight to point O, m.

Example 6-2:

A foundation pit with an excavation depth of $H=5.5$ m is planned to adopt agravity cement-soil wall. The embedded depth of gravity cement-soil wall is 6.5 m, and its unit weight is 19 kN/m³. Figure 6-19 shows the distribution of the active and passive soil pressure strength on both sides of the wall (unit: kPa). Please calculate the width of the gravity cement-soil wall that meets the overturning stability requirements.

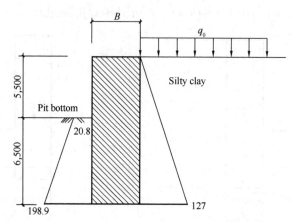

Figure 6-19 distribution of the active and passive soil pressure strength on both sides of the wall (unit: kPa)

Solution:

According to Equation 6-15 $\dfrac{E_{pk}a_p + (G - u_m B)a_G}{E_{ak}a_a} \geq 1.3$

$$\dfrac{20.8 \times 6.5 \times \dfrac{6.5}{2} + \dfrac{1}{2} \times (198.9 - 20.8) \times 6.5 \times \dfrac{6.5}{3} + (6.5 + 5.5) \times B \times 0.5B \times 19}{\dfrac{1}{2} \times 127 \times (6.5 + 5.5) \times \dfrac{(6.5 + 5.5)}{3}}$$

$\geq 1.3 \Rightarrow B \geq 4.46$ m

6.4.3 Stability of soil nailing wall

Soil nailing wall is actually a kind of geotechnical reinforcement. By using the different modulus between steel and soil, friction is generated at their interface, thus restraining the soil and improving its strength and stiffness. The reinforced composite soil forms a gravity retaining wall, so the overall stability of the soil nailing wall is similar to that of the cement soil wall. The anti-sliding stability, anti-overturning stability, pit bottom uplift and overall arc sliding stability should also be carried out. In addition, local stability and anchor stability of soil nails should be satisfied in soil nailing wall.

In order to meet the above stability requirements, soil nailing walls should generally meet the following conditions in design and construction. The wall surface of the soil nailing wall forms an inclination angle of 0°-25° with the vertical direction. The horizontal angle of soil nails is generally 5°-20°; The length of soil nails should not be less than 6 m. The horizontal spacing of soil nails is (10-15)d with d of the diameter of anchor body (steel bar+ cement mortar), and the general horizontal and vertical spacing is 1.0-2.0 m. The soil nail adopts threaded steel bar with a diameter of not less than 16 mm and grade not less than HRB 400. The steel bar is grouted using cement mortar or cement slurry with its strength should not be less than 20 MPa.

1. Overall stability of soil nailing wall

In recent years, the soil nailing wall often adopts different length of soil nails. Longer soil nails are used for the composite soil nailing wall. Thus, the overall overturning and sliding stability of soil nailing wall is no longer checked. The circular sliding surfaces with different sliding surfaces are conducted by using Swedish Slice Method, as shown in Figure 6-20.

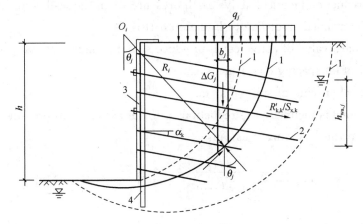

Figure 6-20 Overall stability of soil nailing wall
1 - sliding surface; 2 - soil nail or anchor rod; 3 - sprayed concrete surface layer; 4 - cement soil piles or micro-pile

Because the circular sliding surface passes through some soil nails and anchors, the anti-sliding moment generated by them should be included. The calculation formula is the same as Equation 6-9. The ultimate tension value $R'_{k,k}$ of soil nail or anchor rod in the kth layer takes the smaller of the following two values:

(1) The standard value of ultimate anchor force outside the sliding surface of soil nail or anchor rod;

(2) The standard value of ultimate tensile strength of soil nail or anchor.

The stability analysis of circular sliding should also assume various sliding surfaces with different centers and radii, and the minimum safety factor should be higher than K_s.

2. Anti-pulling stability of soil nails

The anti-pulling stability of a single soil nail should meet the requirements given in Equation 6-16.

$$\frac{R_{k,j}}{N_{k,j}} \geqslant K_t \tag{6-16}$$

where

K_t —— anti-pulling safety factor of soil nail. The K_t is 1.6 and 1.4 corresponding to the safety grade of B and C;

$N_{k,j}$ —— the axial tension of the jth layer soil nail, kN. It is calculated according to Equation 6-17, kN;

$R_{k,j}$ —— the ultimate axial tension capacity of the jth layer soil nail, kN.

$$N_{k,j} = \frac{1}{\cos\alpha_j} \xi \eta_j p_{ak,j} s_{x,j} s_{z,j} \tag{6-17}$$

where

α_j —— the inclination angle of the jth layer soil nail, °;

ξ —— the reduction factor of active earth pressure when the wall is inclined. It should be determined according to Equation 6-18;

$p_{ak,j}$ —— standard value of active earth pressure at the jth layer soil nail, kPa;

$s_{x,j}$ —— horizontal spacing of soil nails, m;

$s_{z,j}$ —— vertical spacing of soil nails, m.

When the slope is inclined, the reduction factor of active earth pressure can be calculated according to the following equation:

$$\xi = \tan\frac{\beta - \varphi_m}{2} \left[\frac{1}{\tan\frac{\beta + \varphi_m}{2}} - \frac{1}{\tan\beta} \right] / \tan^2\left(45° - \frac{\varphi_m}{2}\right) \tag{6-18}$$

where

β —— the included angle between slope surface of soil nailing wall and horizontal plane, °;

φ_m —— average value of equivalent internal friction angle of each soil layer above the excavation bottom weighted by soil layer thickness, °.

It is worth noting that in the soil nailing wall, the tension distribution of the actual soil nail does not completely conform to the linear distribution of Rankine earth pressure. Due to the constraint of boundary conditions, the maximum tensile force of each row of soil nails does not necessarily occur at the same time when the pit is excavated to the bottom. Thus, the tensile force distribution of soil nails is adjusted.

The anti-pulling capacity of a single soil nail can be calculated according to the Equation 6-19, but it should be verified using soil nail anti-pulling tests.

$$R_{k,j} = \pi d_j \sum q_{sik} l_i \tag{6-19}$$

where

d_j —— the diameter of the jth layer soil nail, m;

q_{sik} —— the ultimate bond strength of the jth nail in the ith layer soil, kPa; Table 6-4 gives the reference value;

l_i —— the length of the jth layer soil nail outside the sliding surface, m. When calculating the ultimate anti-pulling capacity of a single soil nail, the straight sliding surface shown in Figure 6-21 can be taken. The angle between the straight sliding surface and the horizontal plane can be taken to $\dfrac{\beta+\varphi_m}{2}$.

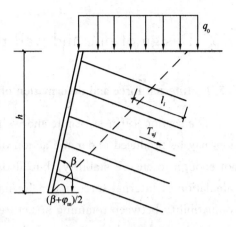

Figure 6-21 Anti-pulling capacity of soil nails

The ultimate bond strength of soil nail q_{sjk} Table 6-4

Soil type	State of soil	q_{sik}/kPa	
		Borehole grouting soil nails	Driving steel pipe soil nails
Plain fill		15-30	20-35
Mucky soil		10-20	15-25
Cohesive soil	$0.75 < I_L \leqslant 1.00$	20-30	20-40
	$0.25 < I_L \leqslant 0.75$	30-45	40-55
	$0 < I_L \leqslant 0.25$	45-60	55-70
	$I_L \leqslant 0$	60-70	70-80
Silt		40-80	50-90
Sand	Loose	35-50	50-65
	Slightly dense	50-65	65-80
	Medium dense	65-80	80-100
	Dense	80-100	100-120

3. Tensile strength of soil nails

The tensile strength of soil nailing rod body shall meet the following requirements:

$$N_j \leqslant f_y A_s \quad (6\text{-}20)$$

where

N_j —— axial tension of the jth layer soil nail, kN;

f_y —— tensile strength design value of soil nailing rod, kPa;

A_s —— cross-sectional area of soil nailing rod, m².

6.5 Design of pile and wall retaining structure

6.5.1 Internal force and deformation of pile or wall retaining structure

The section stiffness of pile and wall retaining structure is small. Significant deformation may be produced under the action of lateral water and soil pressure. Therefore, it is not enough to ensure stability, but also to calculate internal force and deformation. The calculation of internal force and deformation is based on the interaction and deformation compatibility between retaining structure and subgrade soil. The simplest interaction calculation is the lateral elastic foundation reaction method, or the lateral elastic foundation beam method. It is the same as the principle of Wencker foundation beam. Because this beam is placed vertically, the foundation resistance coefficient (subgrade coefficient) is no longer constant. It is related to the depth, that is, the m method is adopted. For 2D problems, the lateral elastic foundation reaction method can generally be used to calculate the retaining structure per unit width, including the force of pile or wall per unit width, internal support (or anchor rod) per unit width and the soil behind the wall per unit width. The internal support (or anchor) is adopted as elastic support. The retaining structure is adopted as elastic beam, and the foundation soil below the pit bottom is adopted as elastic foundation. The internal force and deformation of the structure are analyzed according to the deformation compatibility, as shown in Figure 6-22.

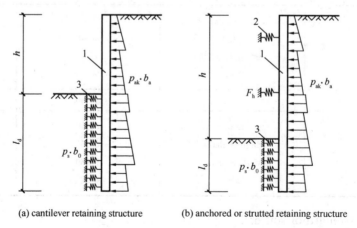

(a) cantilever retaining structure   (b) anchored or strutted retaining structure

Figure 6-22 Lateral elastic foundation reaction method
1-retaining structure; 2-anchor or strut; 3-elastic foundation

The following differential equation of elastic beam can be used for the force and deformation calculation of retaining structure:

$$EI \frac{d^4 v}{dz^4} = p_a + \Delta p_w - k_s v \qquad (6\text{-}21)$$

where

E —— elastic modulus of retaining structure beam material, kN/m^2;

I —— inertia moment of beam section, m^4;

p_a —— the lateral earth pressure in Figure 6-19, take the active and passive earth pressure difference on both sides below the excavation surface, kPa;

Δp_w —— water pressure difference on both sides, kPa;

k_s —— subgrade resistance coefficient of, kN/m^3;

v —— the deflection of the beam, that is, the x displacement in the direction, m.

The subgrade resistance coefficient increases linearly with the depth, which is calculated by using Equation 6-22:

$$k_s = m(z-h) \qquad (6-22)$$

where

m —— the proportional coefficient of subgrade resistance coefficient, kN/m^4;

z —— the distance between the calculation point and the ground surface, $z > h$, m.

For the anchored or strutted retaining structure, the horizontal reaction force of elastic support per unit width is calculated by using Equation 6-23.

$$F_h = k_R(v_R - v_{R0}) + P_h \qquad (6-23)$$

where

F_h —— horizontal reaction force of elastic support per unit width, kN/m;

k_R —— stiffness coefficient of elastic support per unit width, (kN/m)/m;

v_R —— horizontal displacement at support point of retaining structure, m;

v_{R0} —— initial horizontal displacement of support point when setting support or anchor, m;

P_h —— pre-tension in horizontal direction per unit width, kN/m.

Among them, the stiffness coefficient of elastic support k_R is relatively easy to determine. For the anchor, it can be determined by field pull-out test, as shown in Equation 6-24:

$$k_R = \frac{Q_2 - Q_1}{(s_2 - s_1)s} \qquad (6-24)$$

where

Q_1, Q_2 —— the load value corresponding to the locking value and the standard axial tension of anchor rod on the Q-s curve by doing cyclic loading or step-by-step loading tests, kN;

s_1, s_2 —— the anchor head displacement corresponding to the load Q_1 and Q_2 on the Q-s curve, m;

s —— horizontal spacing of anchor rod, m.

For the strut, the stiffness coefficient of elastic support can be obtained by linear elastic calculation of bar.

6.5.2 Anchor and strut support

Compared with the cantilever retaining structure, the use of anchor and strut support can not only greatly reduce the internal force on the retaining structure, but also reduce the deformation of the retaining structure and the settlement of the surrounding ground.

1. Soil anchor

For the site with good soil conditions and soil anchor construction conditions, soil anchor should be considered first. The anchored retaining structure can provide an open space in the excavation, which is convenient for excavation and construction. Thus, the construction period can be shortened. The soil anchor is composed of anchor head, anchor rod and anchor body, as shown in Figure 6-6. The calculation of anchor rod is similar to that of soil nail. However, there are some differences between anchor rod and soil nail due to the prestress of anchor rod and pressure grouting of anchor body.

(1) Anti-pulling resistance calculation of anchor rod

The anti-pulling resistance calculation of anchor rod is similar to soil nail. It can be expressed as Equation 6-25:

$$\frac{R_K}{N_K} \geqslant K_t \tag{6-25}$$

where

K_t —— anti-pulling safety factor of anchor rod. The K_t is 1.8, 1.6 and 1.4 corresponding to the safety grade of A, B and C;

N_K —— axial tension of anchor rod, kN, calculated according to Equation 6-26;

R_K —— ultimate anti-pulling capacity of anchor rod, kN. It is determined according to Equation 6-27.

$$N_K = \frac{F_h s}{b_a \cos\alpha} \tag{6-26}$$

where

F_h —— the horizontal reaction force of elastic support per unit width of retaining structure, kN/m, calculated according to Equation 6-23;

s —— horizontal spacing of anchor rod, m;

b_a —— calculated width of the retaining structure, m;

α —— angle of anchor rod, °.

The ultimate anti-pulling capacity of anchor rod should be determined by anti-pulling test. It can also be estimated according to Equation 6-27, but it should be verified by anti-pulling test.

$$R_K = \pi d \sum q_{sik} l_i \tag{6-27}$$

where

d —— diameter of anchor body, m;

6.5 Design of pile and wall retaining structure

l_i —— the anchoring section length of the anchor rod in the ith soil layer, m. It is the length of the anchor rod outside the theoretical linear sliding surface. The theoretical linear sliding surface is determined according to Figure 6-23;

q_{sik} —— the ultimate bond strength between the anchor body and the ith soil layer, kPa. It should be taken according to the engineering experience and combined with Table 6-5.

Standard value of ultimate bond strength of anchor body q_{sik} Table 6-5

Name of soil	Soil state or density	q_{sik}/kPa	
		Primary constant pressure grouting	Secondary pressure grouting
Filling soil		16-30	30-45
Mucky soil		16-20	20-30
Cohesive soil	$I_L > 1$	18-30	25-45
	$0.75 < I_L \leqslant 1.00$	30-40	45-60
	$0.50 < I_L \leqslant 0.75$	40-53	60-70
	$0.25 < I_L \leqslant 0.50$	53-65	70-85
	$0 < I_L \leqslant 0.25$	65-73	85-100
	$I_L \leqslant 0$	73-90	100-130
Silt	$e > 0.90$	22-44	40-60
	$0.75 \leqslant e \leqslant 0.9$	44-64	60-90
	$e < 0.75$	64-100	80-130
Fine silty sand	Slightly dense	22-42	40-70
	Median dense	42-63	75-110
	Dense	63-85	90-130
Medium sand	Slightly dense	54-74	70-100
	Median dense	74-90	100-130
	Dense	90-120	130-170
Coarse sand	Slightly dense	80-130	100-140
	Median dense	130-170	170-220
	Dense	170-220	220-250
Gravelly sand	Medium-dense and dense	190-260	240-290
Weathered rock	Completely weathered	80-100	120-150
	Strong weathered	150-200	200-260

Note: 1. When the content of fine particles in sandy soil exceeds 30% of the total mass, the values in the table should be multiplied by the coefficient of 0.75.

2. For organic soil with organic matter content of 5%-10%, the valuesin the table should be appropriately reduced.

3. When the length of the anchor section is greater than 16 m, the values in the table shall be appropriately reduced.

The length of the free segment of the anchor rod l_f shall be determined according to Equation 6-28. It should be longer than 5.0 m and 1.5 m outside the sliding surface.

$$l_f \geqslant \frac{(a_1 + a_2 - d\tan\alpha)\sin\left(45° - \frac{\varphi_m}{2}\right)}{\sin\left(45° + \frac{\varphi_m}{2} + \alpha\right)} + \frac{d}{\cos\alpha} + 1.5 \qquad (6-28)$$

where

l_f —— length of free segment of anchor rod, m;

α —— inclination angle of anchor rod, °;

a_1 —— distance from the midpoint of anchor head of anchor rod to the pit bottom, m;

a_2 —— the distance from the pit bottom to the point O, where the earth pressure on both sides (active and passive) is equal, m;

d —— horizontal dimension of retaining structure, m;

φ_m —— average value of equivalent internal friction angle weighted by thickness of each soil layer above point O, °.

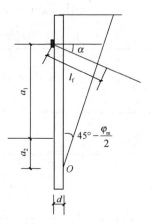

Figure 6-23 Theoretical linear sliding surface

(2) Tensile strength verification of anchor rod

The tensile strength verification of the anchor rod shall be conducted according to Equation 6-29:

$$N \leqslant f_{py} A_p \qquad (6-29)$$

where

N —— design value of axial tension of anchor rod under basic combination of loadings, kN;

f_{py} —— design value of tensile strength of prestressed reinforcement, kPa;

A_p —— cross-sectional area of prestressed reinforcement, m².

2. Internal strut support

Soil anchor is not applicable in the following situations: 1) The soil layer is soft and cannot provide enough anchoring force for the anchor; 2) There are underground structures and important public facilities closely adjacent to the pit; 3) The space under the adjacent building foundation are not allowed to set the anchor section of the anchor. In these cases, it is necessary to set internal strut support inside the excavation.

As Figures 6-24 and 6-25 shown, two types of internal strut support are widely used: steel support and cast-in-situ reinforced concrete support.

(1) Steel support

The steel support can use single steel pipe, double steel pipe, single I-beam, double I-beam, H-beam, channel steel, and their combination. Horizontal bracing and slant support are arranged vertically. Their general layout can be diagonal braces, well-shaped

6.5 Design of pile and wall retaining structure

Figure 6-24 Steel support

Figure 6-25 Cast-in-situ reinforced concrete support

braces, and angle braces. It can also be used in conjunction with reinforced concrete support, but deformation compatibility should be handled carefully. The steel support has the advantages of convenient installation and demolition, reusability, pre-stressed support, adjustable axial force to effectively control the deformation of the retaining wall. However, its construction requirements are high. For example, improper handling of nodes and supporting structures may cause instability of the support.

(2) Cast-in-situ reinforced concrete support

For cast-in-situ reinforced concrete support, its shape and size can be determined according to the requirements. The cast-in-situ reinforced concrete support has high stiffness, small deformation, high reliability of strength, and convenient construction. However, the pouring and curing time is long. The retaining structure is in an unsupported state for a long time, and the displacement in soft soil is large. If there are high requirements for controlling deformation, it is necessary to reinforce the soft soil in the passive area. Meanwhile, long construction period and difficult demolition has an impact on the surrounding environment.

Questions

6-1 What types of support structures are there? What conditions do they apply to?

6-2 What are the applicable conditions for slope excavation?

6-3 Please explain the mechanism and applicable conditions of using soil nail walls to reinforce slopes.

6-4 What are the similarities and differences between soil nails and anchor rods in reinforcement mechanisms, construction methods and design calculations?

6-5 What is a reverse approach? What are the applicable conditions?

6-6 How should the soil pressure and water pressure on the support structure be calculated below the water level?

6-7 What should be included in the stability analysis of soil nail walls?

6-8 In the support structure, how many parts are the anchor rods divided along their length? How to determine the length of each part?

6-9 There is a uniform cohesive soil subgrade. Its unit weight γ is 19 kN/m^3, cohesion c is 15 kPa, and internal friction angle φ is 26°. A foundation pit is excavated in this uniform cohesive soil subgrade with a excavation depth of 15 m. A uniformly distributed load of 48 kPa is applied on the ground outside the pit. Please determine the depth where the active and passive earth pressure strength are equal.

6-10 A strip foundation pit is excavated in saturated soft clay, which uses 11 m long cantilever steel sheet piles for support. The pile top is at the same height with the ground. The unit weight of soft soil is 17.8 kN/m^3, and the shear strength of soil measured using vane shear test τ is 40 kPa. The load applied on the ground is 10 kPa. In order to meet the uplift stability requirements of the soil at the bottom of the steel sheet piles, try to determine the maximum excavation depth of this foundation pit.

Chapter 7 Ground Improvement

7.1 Introduction

Ground improvement, or ground modification, is defined as the alteration of site foundation soils or project earth structures to provide better performance under design and/or operational loading conditions. Ground improvement is necessary when poor soil conditions are encountered for the purpose at hand. While the poor soil conditions could readily be dealt with by excavating and replacing the soil, or perhaps by using deep foundations, it is often more cost effective to simply improve the soil in place through some type of treatment.

7.1.1 Type and characteristics of soft soil

The soft soil which needs to be treated generally mainly includes silt and muddy soil, loose sand, flushing fill, miscellaneous fill, peat soil and other high compressibility soils. Sometimes some special soils, such as expansive soil and collapsible loess, should be treated according to their characteristics.

1. Muck and mucky soil

Muck and mucky soil refer to the soil deposited in still water or very slow flowing water environment in the late Quaternary, and the natural water content less than the liquid limit and the porosity ratio greater than or equal to 1.0. Among them, when the natural porosity ratio e is greater than or equal to 1.5, it is called muck. When the void ratio e is 1.0-1.5, it is called mucky soil.

Muck and mucky soil have the following characteristics.

(1) High compressibility with an average compressibility of 3×10^{-3}- 5×10^{-4} kPa^{-1}.

(2) The shear strength is low, the undrained strength is 10-20 kPa, the standard penetration test blow number is less than 5, and the subgrade bearing capacity is less than 100 kPa.

(3) The permeability is small, and the permeability coefficient is generally 1×10^{-8}-1×10^{-10} m/s.

(4) It has obvious thixotropy and rheology.

2. Flushing fill

Flushing fill refers to the sedimentary soil formed by hydraulic flushing and filling of mud and sand at the bottom of rivers and ports with dredgers or mud pumps, also known as dredged fill. The composition of filling soil is complex, and most of them belong to cohesive soil, silt or silt. This kind of soil has high water content, which is often greater than the liquid limit. Among them, the filled soil with more clay particles has slow drainage consolidation, and most of it belongs to under-consolidated soil with high compressibility and low strength. The mechanical properties of flushing soil are worse than those of similar natural soils.

3. Miscellaneous fill

Miscellaneous fill refers to the waste accumulation formed by human activities without careful compaction. It includes industrial waste, construction waste and domestic waste. The composition of miscellaneous fill is complex and its distribution is irregular. The properties of miscellaneous fill change with the age of landfill. It is generally believed that the properties will gradually become stable when the age of landfill is more than five years. In addition, miscellaneous fill often contains humus and hydrate, especially containing domestic garbage with higher humus content. With the decay of organic matter, the settlement of foundation will increase and be uneven. Thus, the bearing capacity and compressibility of different positions in the same site will often be quite different.

4. Loose sand

Loose sand refers to sand whose relative density is less than or equal to 1/3 or whose standard penetration test blow number is no more than 10. Saturated loose sand is easy to flow and slip under the action of small shear stress, and sand flow occurs. In addition, saturated loose sand is easily liquefied by vibration.

5. Peat soil

The inorganic soil is defined as soil with its organic matter content W_u less than 5%. If $5\% \leqslant W_u \leqslant 10\%$, the soil is called organic soil. The peaty soil refers to the soil with $10\% < W_u \leqslant 60\%$ and the peat soil refers to the soil with $W_u > 60\%$. The peaty soil and peat soil are usually formed in low-lying swamps and shrub belts and they are often in a saturated state with water content as high as several hundred percent. Thus, they usually have low density and high great compressibility. Because the content and decomposition degree of plants are different, the properties of this kind of soil are very uneven, which easily leads to serious uneven deformation of buildings. Meanwhile, the bearing capacity of this kind of soil is very low, which belongs to the soil with the worst properties and is generally not suitable as the foundation of buildings.

A subgrade consisting of or mainly composed of the above types of soil is called a soft subgrade. Whether the ground improvement is needed is not only related to the weakness of the subgrade, but also related to the nature of the building. In the case that buildings are very important and require high stability and small deformation, it may be required to treat the subgrade, even if the subgrade soil is not very weak. On the contrary, in the case that the importance of buildings is low and the requirements for subgrade are not high, it may not be necessary to carry out ground treatment, even if the subgrade soil is weak. Therefore, ground improvement is a complex problem that needs to comprehensively consider soil and buildings.

7.1.2 Purpose and requirements of ground improvement

Ground improvement typically serves one or more of the following primary functions:
- Increase shear strength and bearing capacity;
- Increase density;
- Decrease permeability;
- Control deformations;
- Increase drainage;
- Accelerate consolidation;
- Decrease imposed loads;
- Provide lateral stability;
- Increase resistance to liquefaction;
- Transfer embankment loads to more competent subsurface layers.

Soil improvement has various benefits as follows:
- Less consumption of time as the design and implementation is relatively fast;
- When the techniques are applied for ground improvement, it hardly produces any waste. Thus, no disposal costs are engaged in it;
- Easy design and construction of substructure;
- Ground improvement is applicable and also effective on various types of soils.

The improved ground must meet the requirements of subgrade bearing capacity, deformation and stability according to Chapter 3 of this book. Because the ground improvement belongs to local treatment, the width and depth correction coefficient of bearing capacity is different from the naturally deposited soil layer. According to the *Technical Specification for Building Ground Improvement* (*JGJ 79—2012*), despite large area compacted fill, the width correction coefficient of bearing capacity of all treated foundations is zero, and the depth correction coefficient is 1.0.

7.1.3 Classification of ground improvement methods

These are various techniques available for ground improvement. The methods of ground improvement can be categorized based on the types of soil, treatment time, and mechanism of ground treatment. According to the different mechanism of ground treatment, the classification of ground improvement methods can be given in the Table 7-1.

Chapter 7 Ground Improvement

Classification of ground improvement methods   Table 7-1

Type	Method	Principle	Applicability
Replacement	Replacement cushion method	Dig out the soft soil layer or uneven soil layer in a certain range under the foundation bottom surface and backfill other materials with stable performance, no corrosiveness and high strength, and compact the cushion layer	Suitable for shallow treatment of soft or unfavorable foundations such as silt, muddy soil, plain fill, miscellaneous fill, collapsible loess, expansive soil, seasonally frozen soil, and etc.
	Sand-gravel columns	By using vibration technology or high-pressure water flushing technology, holes are formed in soft soil layers, and then coarse materials such as crushed stones are backfilled to form pile columns. They are combined with the original foundation soil to form a composite foundation to improve the bearing capacity of the foundation and reduce settlement	Suitable for silt, muddy soil, silt, cohesive soil, and artificial fill with undrained shear strength greater than 20 kPa
	Lime piles	Using Luoyang shovels or machinery to form holes, the mixture of quicklime and fly ash is evenly mixed, layered and compacted in the hole to form a vertical reinforcement. The combination of the soil and piles forms a composite foundation	Suitable for saturated cohesive soil, silt, muddy soil, plain fill, and miscellaneous fill subgrades
	CFG piles	By using vibration sinking and long spiral drilling to form holes, backfilling with a mixture of cement, fly ash, crushed stone, or sand to form a high cohesive strength pile. The piles, soil between piles, and a cushion layer forms a composite foundation	Suitable for cohesive soil, silt, sandy soil, and self-weight consolidated plain fill soil
	Dynamic replacement	Lift the heavy hammer to a high place and let it fall freely to form a compaction pit, and continuously compact the hard granular materials such as sand, stone, and steel slag filled in the pit to form a dense pier body	Suitable for high saturation silt, soft to plastic cohesive soil
Drainage consolidation	Surcharge preloading method Vacuum preloading method	Before constructing a structure, the drainage conditions of the foundation are improved by adding vertical or horizontal drainage bodies, and preloading loads are applied to the foundation through measures such assurcharge or vacuum to accelerate the drainage consolidation and strength growth of the foundation soil	Suitable for thick saturated cohesive soil foundations such as silt, muddy soil and fill soil

7.1 Introduction

continued

Type	Method	Principle	Applicability
Deep compaction	Dynamic compaction	Using a heavy rammer to fall freely from a high place, the foundation soil is compacted under the impact and vibration forces of dynamic compaction	Suitable for crushed stone soil, sandy soil, low saturation silt and cohesive soil, collapsible loess, plain fill and miscellaneous fill
	Sand-gravel compaction pile	By using vibration sinking or impact, holes are formed in the foundation. Sand piles, gravel piles, and etc. are set up to produce vibration compaction and compaction effects on the surrounding soil during the pile making process. The piles and the soil between piles form composite foundation	Suitable for compacting loose sand, silt, cohesive soil, plain fill, miscellaneous fill and other foundations
	Compacted soil-lime columns or soil columns	Utilize lateral extrusion drilling equipment to create holes and compact the soil between piles. Using soil and lime or plain soil into the pile hole and compact it layer by layer to form the pile body. The piles and the soil between piles form composite foundation	Suitable for collapsible loess, plain fill, and miscellaneous fill above groundwater level
	Vibro-compaction	Under the combined action of horizontal vibration of the vibro-flotation device and high-pressure water, the loose sand layer is vibrated and compacted, thereby improving the bearing capacity of the foundation, reducing settlement, and improving the liquefaction resistance of the foundation soil. This method can add backfill or not, and adding backfill is also known as the vibro-flotation compaction gravel pile method	Suitable for saturated and loose sandy soil foundations
Chemical grouting	Grouting method	By injecting cement slurry or other chemical slurry, or spraying or mechanical mixing cement slurry, soil particles are cemented to improve the foundation	Suitable for rock foundation, sandy soil, silty soil, cohesive soil, and general fill soil
	Cement deep mixing method		Suitable for silt and muddy soil, silt, saturated loess, plain fill, cohesive soil, and saturated loose sand without flowing groundwater
	High pressure jet grouting		Suitable for silt, muddy soil, flow plastic or soft plastic or plastic cohesive soil, silty soil, sandy soil, loess, plain fill, and gravel soil

continued

Type	Method	Principle	Applicability
Others	Reinforcement method	Lay reinforced materials (such as geotextile, geogrid, metal strip, soil nail, etc.) in the foundation to form a reinforced soil cushion, so as to increase the pressure diffusion angle and improve the bearing capacity and stability of the foundation	Suitable for manually filled embankments, retaining wall structures, and soil slope reinforcement and stability
	Freezing method	Freeze the soil, improve the water interception performance of the foundation soil, increase the shear strength of the soil, and form a retaining structure or water stop curtain	Saturated sand or soft clay as temporary construction measures
	Roasting method	Borehole heating or roasting can reduce soil moisture content, reduce compressibility and improve soil strength	Soft clay and collapsible loess, suitable for areas with sufficient heat sources
	Correction method	By using methods such as loading, soil excavation, and jacking to correct uneven ground settlement, the goal of correction is achieved	Various types of poor foundations

7.2 Replacement cushion method

The replacement cushion method is to remove all or part of the soft soil layer or uneven soil layer within a certain depth under the ground, and then backfill the materials with high strength and stable performance such as sand, gravel, plain soil, lime soil, fly ash and etc. in layers.

When the bearing capacity and deformation of soft foundation cannot meet the requirements of buildings, and the thickness of soft soil layer is not very large, the soil replacement cushion method is a relatively economical and simple shallow treatment method for soft soil foundation. Different backfill materials form different cushions, such as sand cushion, crushed stone cushion, plain soil or lime soil cushion, fly ash cushion and cinder cushion.

The replacement method is suitable for shallow treatment of silt, mucky soil, collapsible loess, plain fill, miscellaneous fill foundation, culvert and hidden pond. It is often used for ground improvement of light buildings, floors, stacking sites and road engineering. When the building load is small and the thickness of soft soil layer is small, the replacement cushion method can achieve good results. The thickness of replacement cushion should be determined according to the depth of replacement soft soil and the bearing capacity of the underlying layer, and the thickness should be 0.5-3 m.

7.2.1 Function of replacement cushion

The function of soil replacement cushion in soft soil subgrade can be given in the following aspects.

1. Improve the bearing capacity of shallow foundation

Replacing soft soil with sand or other filling materials with high shear strength can improve the bearing capacity of foundation, and spread the foundation pressure to the soft soil below the cushion to avoid foundation damage.

2. Reduce the deformation of foundation

Generally, the settlement of shallow part of foundation accounts for a relatively large proportion of the total settlement. Taking strip foundation as an example, the settlement in the depth range equivalent to the width of foundation accounts for about 50% of the total settlement. If the upper soft soil layer is replaced by dense sand or other filling materials, the settlement of this part can be reduced. Due to the stress diffusion of sand cushion or other cushion, the pressure acting on the underlying soil is small, which will correspondingly reduce the settlement of the underlying soil.

3. Accelerate the drainage consolidation of soft soil layer

Cushion materials such as sand cushion and gravel cushion have high water permeability. After the soft soil is compressed, the cushion can be used as a good drainage surface. It can quickly dissipate the pore water pressure under the foundation, accelerate the consolidation of the soft soil under the cushion, improve its strength and avoid the plastic failure of the foundation soil. Using permeable material as cushion is equivalent to adding a layer of horizontal drainage channel, which plays a drainage role. In the process of building construction, the dissipation of pore pressure is accelerated, and the increase of effective stress is also accelerated, which is conducive to improving the bearing capacity of foundation, increasing the stability of foundation, accelerating the construction progress and reducing the post-construction settlement of buildings.

4. Prevent frost heaving of soil

Because the coarse-grained cushion material has large pores, it is not easy to produce capillary phenomenon. Thus, it can prevent frost heaving caused by freezing in soil in cold areas. At this time, the bottom of sand cushion should meet the requirements of local freezing depth.

5. Eliminate the collapsibility, expansion and contraction of special soil

For special soils such as collapsible loess, expansive soil or seasonal frozen soil, the main purpose of treatment is to eliminate or partially eliminate the collapsibility and expansion and contraction of foundation soil. Materials such as sand, gravel, rubble, cinder, lime-fly ash or lime soil can be used as cushion to eliminate expansion and contraction, but the cushion thickness should be determined according to deformation calculation. Generally, cushion thickness is not less than 0.3 m, and its width should be greater than that of foundation. Both sides of foundation should be backfilled with the same materials as cushion.

7.2.2 Design of replacement cushion

The design of cushion should not only meet the requirements of buildings for foundation deformation and stability, but alsoshould be economical and rational. When designing, comprehensive analysis should be made according to the type, structural characteristics, loading properties, geotechnical engineering conditions, construction machinery and equipment, and the properties and source of fillers. Its main design is to determine the reasonable thickness and width of replacement. Figure 7-1 shows the schematic diagram of replacement cushion method.

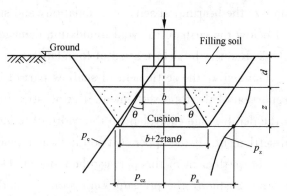

Figure 7-1 Schematic diagram of replacement cushion method

1. Determination of cushion thickness z

The cushion thickness z should be determined according to the bearing capacity of the underlying soft soil layer at the bottom of the cushion. The internal stress distribution is also shown in Figure 7-1. It should meet the requirements of Equation 7-1:

$$p_z + p_{cz} \leqslant f_{az} \tag{7-1}$$

where

p_z——the additional stress at the bottom of cushion corresponding to the standard combination of load effect, kPa;

p_{cz}——self-weight stress of soil at the bottom of cushion, kPa;

f_{az}——characteristic value of bearing capacity at the bottom of cushion layer after depth correction, kPa.

The additional stress at the bottom of cushion can be calculated according to the pressure diffusion angle θ.

Strip foundation

$$p_z = \frac{b(p_k - p_c)}{b + 2z\tan\theta} \tag{7-2}$$

Rectangular foundation $\quad p_z = \dfrac{lb(p_k - p_c)}{(l + 2z\tan\theta)(b + 2z\tan\theta)} \tag{7-3}$

where

b——width of bottom surface of strip foundation or rectangular foundation, m;

l——length of bottom surface of rectangular foundation, m;

p_k——average pressure value at the bottom of foundation corresponding to the standard combination of load effect, kPa;

p_c——self-weight stress of soil at the bottom of foundation, kPa;

z——thickness of the cushion below the foundation bottom, m;

θ——pressure diffusion angle of cushion, °, which is given in Table 7-2.

Cushion pressure diffusion angle θ   Table 7-2

z/b \ Replacement material	Medium sand, coarse sand, gravel sand, round gravel, breccia, stone chips, pebbles, gravel and slag	Silty clay and fly ash	Lime soil
0.25	20°	6°	28°
≥0.50	30°	23°	

Note: 1. When z/b is less than 0.25, $\theta = 0°$ is used for all materials except lime soil, which should be determined by experiments if necessary.
2. When $0.25 < z/b < 0.50$, the θ can be determined by interpolation.
3. The θ of geosynthetics reinforced cushion should be determined by field test.

When calculating, estimate a cushion thickness first, and then use Equation 7-1 to verify. If it does not meet the requirements, increase the thickness and recalculate until it meets the requirements. The cushion thickness should not be less than 0.5 m or more than 3.0 m. Too thick is uneconomical to construct.

2. Determination of cushion width

The width b' of cushion should not only meet the stress diffusion requirements, but also prevent the cushion from extruding to both sides. It can be calculated according to Equation 7-4.

$$b' \geq b + 2z\tan\theta \tag{7-4}$$

where

b' —— cushion width at the bottom, m.

After the bottom width of the cushion is determined, it will extend to the ground according to the slope required for foundation pit excavation. The design section of the cushion will be obtained. Generally, the width of the top surface of the cushion should not be less than 0.3 m beyond the bottom edge of the foundation. The cushion width can be appropriately widened according to the construction requirements.

3. Determination of cushion bearing capacity

The bearing capacity of cushion should be determined by field test. When there is no test data, it can be selected according to Table 7-3, and the bearing capacity of substratum should be verified.

Bearing capacity and compaction requirement of various cushions　　　Table 7-3

Construction method	Types of replacement materials	Compaction coefficient λ_c	Standard value of bearing capacity/kPa
Compaction, vibration and compaction	Gravel and pebble	$\geqslant 0.97$	200-300
	Sand with stones (including gravel and pebbles accounting for 30%-50% of the total weight)		200-250
	Soil with stones (including gravel and pebbles accounting for 30%-50% of the total weight)		150-200
	Medium sand, coarse sand, gravel sand, breccia, round gravel and stone chips		150-200
	Silty clay		130-180
	Lime soil	$\geqslant 0.95$	200-250
	Fly ash	$\geqslant 0.95$	120-150

Note: 1. The compaction coefficient λ_c is the ratio of the controlled dry density ρ_d of soil to the maximum dry density ρ_{dmax}. The ρ_{dmax} should be determined by compaction test, and the maximum dry density of gravel or pebble should be $(2.1-2.2) \times 10^3$ kg/m³.

2. The compaction coefficient λ_c in the table is the compaction control standard given when the maximum dry density ρ_{dmax} of soil is measured by light compaction test. The compaction standard for silty clay, lime soil, fly ash and other materials is $\lambda_c \geqslant 0.94$ when heavy compaction test is adopted.

Example 7-1:

The bottom side length of reinforced concrete rectangular foundation under a column is $b \times l = 1.8$ m$\times 2.4$ m, and the buried depth is $d = 1.0$ m. The standard value of axial load applied on the foundation F_k is 800 kN. The first soil layer under the ground is a silty clay layer with a thickness of 0.7 m, and its unit weight γ is 17.0 kN/m³. The muddy soil is under silty clay layer with the unit weight γ is 17.0 kN/m³, and the characteristic bearing capacity of muddy soil f_{ak} is 95 kPa. If coarse sand is used as a sand cushion with a thickness of 1.0 m under the ground, and the cushion unit-weight γ is is 20.0 kN/m³. Try to determine whether the size of foundation bottom surface and the thickness of sand cushion meet the requirements, and determine the size of bottom surface of sand cushion.

Solution:

(1) Cushion thickness calculation

Estimate cushion thickness $z = 1.0$ m, $z/b = 1.0/1.8 > 0.5$, then the pressure diffusion angle $\theta = 30°$.

The self-weight stress at the bottom of the foundation is

$p_c = (17.0 \times 0.7 + 16.5 \times 0.3) = 16.85$ kPa

The base pressure at the bottom of the foundation is

$$p_k = \frac{F_k + G_k}{A} = \frac{800 + 20 \times 1.8 \times 2.4 \times 1}{1.8 \times 2.4} = 205.18 \text{ kPa}$$

The additional pressure at the bottom of the cushion is

$$p_z = \frac{bl(p_k - p_c)}{(b + 2z\tan\theta)(l + 2z\tan\theta)}$$

$$= \frac{1.8 \times 2.4 \times (205.18 - 16.85)}{(1.8 + 2 \times 1 \times \tan 30°)(2.4 + 2 \times 1 \times \tan 30°)} = 77.4 \text{ kPa}$$

The self-weight stress of soil at the bottom of cushion is

$$p_{cz} = (17.0 \times 0.7 + 20.0 \times 1.3) = 37.9 \text{ kPa}$$

According to the table of bearing capacity correction coefficient in Code for Design of Building Foundation, $\eta_d = 1.0$, the characteristic value of bearing capacity of muddy soil at the bottom of cushion after depth correction is as follows:

$$f_{az} = f_{ak} + \eta_d \gamma_{mz}(d + z - 0.5)$$
$$= 95 + 1.0 \times [(17.0 \times 0.7 + 16.5 \times 1.3)/2] \times (2.0 - 0.5) = 120 \text{ kPa}$$
$$p_z + p_{cz} = (77.4 + 37.9) = 115.3 \text{ kPa} < f_{az}$$

It shows that the strength meets the requirements, and the cushion thickness is 1.0 m.

(2) Determine the size of the bottom surface of the cushion layer

$b' \geqslant b + 2z\tan\theta = (1.8 + 2 \times 1 \times \tan 30°) = 2.95$ m , take $b' = 3.0$ m

$l' \geqslant l + 2z\tan\theta = (2.4 + 2 \times 1 \times \tan 30°) = 3.55$ m take $l' = 3.6$ m

7.3 Preloading method

Preloading method is a kind of soft soil ground improvement method. By using the characteristics of foundation drainage consolidation, it applies preloading load and uses various drainage bodies (such as sand drain and drainage cushion). The pore water in the soil is discharged and gradually consolidated. Thus, the subgrade strength is gradually improved. Preloading method is often used to solve the problem of settlement and stability of soft clay subgrade. It can make the settlement of subgrade basically completed or mostly completed during loading and preloading, so that the building will not have excessive settlement and settlement difference during use. At the same time, it can increase the shear strength of subgrade soil, thus improving the bearing capacity and stability of foundation.

According to the different preloading loads, preloading method can be divided into surcharge preloading method, vacuum preloading method and combined preloading method.

7.3.1 Mechanism of various preloading methods

1. Surcharge preloading reinforcement

Preloading method is a method of preloading in the construction site before the building is built, so that the consolidation settlement of foundation is basically completed and the strength of foundation soil is improved.

After the load is applied to the saturated soft soil foundation, the pore water is slowly discharged, and the foundation is consolidated and deformed. At the same time, with the dissipation of hydrostatic pressure and the increase of effective stress, the strength of foundation soil increases gradually. Under load, the consolidation process of soil layer is the process of excess pore water pressure dissipation and effective stress increase. If the total stress increment at a point in the foundation is $\Delta\sigma$, the effective stress increment is $\Delta\sigma'$ and the pore water pressure increment is Δu, they satisfy the following relationship:

$$\Delta\sigma' = \Delta\sigma - \Delta u \qquad (7-5)$$

Preloading the foundation with external loads such as filling soil is a method to increase effective stress $\Delta\sigma'$ by increasing total stress $\Delta\sigma$ and dissipating pore water pressure Δu. Surcharge preloading is drainage consolidation under the condition of excess hydrostatic pressure in foundation, which is called positive pressure consolidation.

The drainage consolidation effect of soil layer is related to its drainage boundary conditions. Figure 7-2 shows the drainage of preloading method. When the thickness of the soil layer is relatively small relative to the load width (or diameter), the pore water in the soil layer is discharged from the upper and lower permeable layers, and the soil layer is consolidated (Figure 7-2a), which is called vertical drainage consolidation. According to consolidation theory, the time required for consolidation of cohesive soil is proportional to the square of drainage distance. Therefore, in order to accelerate the consolidation of soil layer, the most effective method is to increase the drainage channels of soil layer and shorten the drainage distance. Vertical drainage bodies such as sand drains and prefabricated strip drains are set for this purpose, as shown in Figure 7-2(b).

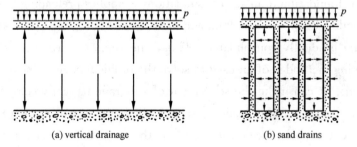

(a) vertical drainage   (b) sand drains

Figure 7-2 Drainage of preloading method

2. Vacuum preloading reinforcement mechanism

Vacuum preloading method is to lay sand cushion on the surface of soft soil foundation, place vertical drainage pipes, and then use airtight sealing membrane to isolate it from the atmosphere. The suction pipe buried in the sand cushion is pumped by vacuum device to form vacuum and increase the effective stress of the foundation. When vacuumizing, negative pressure is gradually formed in the surface sand cushion and the vertical drainage channel, which makes the pressure difference between the soil and the drainage channel and cushion. Under this pressure difference, the pore water in the soil is continuously discharged from the drainage channel, thus consolidating the soil.

The principle of vacuum preloading method can be given in the following.

(1) The membrane bears a load equal to the pressure difference between the inside and outside of the membrane.

(2) When the groundwater level decreases, the additional stress increases accordingly.

(3) The closed air bubbles are discharged, and the permeability of soil increases.

Vacuum preloading is to vacuum under the sealing membrane covering the ground, so that the air pressure difference between the inside and outside of the membrane is formed and the consolidation pressure of the clay layer is generated. That is, under the condition of constant total stress, the effective stress is increased by reducing pore water pressure. Vacuum preloading and dewatering preloading are drainage consolidation under negative hydrostatic pressure, which is called negative pressure consolidation.

In fact, preloading method is a combination of drainage system and pressurization system. If there is no pressurization system, the water in the pores will not be discharged naturally without pressure difference, and the foundation will not be strengthened. If only the consolidation pressure is increased and the drainage distance of the soil layer is not shortened, the settlement required by the design cannot be completed as soon as possible during preloading. The strength cannot be improved in time, and the loading cannot be carried out smoothly. Therefore, the above two systems are always considered together when designing.

Preloading method is suitable for saturated cohesive soil foundations such as silt, muddy soil and filling soil. For cohesive soil with horizontal sand interlayer, because of its good lateral drainage performance, good consolidation effect can be obtained without vertical drainage (sand drain, etc.).

7.3.2 Design of surcharge preloading method

The design of surcharge preloading method includes the following contents:

(1) Select vertical drainage wells and determine their section size, spacing, arrangement and depth;

(2) Determine the preloading area, preloading load, step-loading, loading rate and preloading time;

(3) Calculate the consolidation degree, strength growth, anti-sliding stability and deformation of foundation soil.

1. Vertical drainage well

(1) Diameter of vertical drainage well

Vertical drainage wells are divided into ordinary sand wells, bagged sand wells and prefabricated strip drain. The diameter of ordinary sand wells is 300-500 mm, and that of bagged sand wells is 70-120 mm. Prefabricated strip drain has been standardized with its diameter of 60-70 mm.

(2) Spacing of sand wells or prefabricated strip drains

The spacing between sand wells or prefabricated strip drains can be determined according to the consolidation characteristics of foundation soil and the required consolidation degree within a predetermined time. Usually, the spacing of sand drains can be selected according to the drain spacing ratio n, which is determined by the following equation:

$$n = d_e/d_w \tag{7-6}$$

where

d_e ——effective drainage cylinder diameter of sand drain, mm;

d_w ——diameter of drainage bodies mm.

The spacing of ordinary sand wells can be selected as $n = 6$-8. The spacing between bagged sand wells or prefabricated strip drain can be selected as $n = 15$-22.

Prefabricated strip drain is commonly expressed by equivalent diameter. If the width of prefabricated strip drain is b and the thickness is δ, the equivalent diameter can be calculated according to the following equation:

$$d_p = \frac{2(b+\delta)}{\pi} \tag{7-7}$$

where

d_p —— equivalent diameter of prefabricated strip drain, mm;

b ——width of prefabricated strip drains, mm;

δ ——thickness of prefabricated strip drains, mm.

Figure 7-3 shows the prefabricated strip drains. At present, the size of the prefabricated strip drain used is generally 100 mm×4 mm. It is packaged in rolls, and each roll is a-

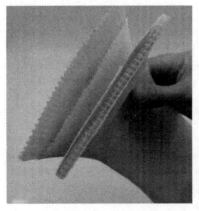

Figure 7-3 Prefabricated strip drains

bout several hundred meters long. A special tape inserting machine is used to insert the prefabricated strip drain into the soft soil foundation.

(3) Arrangement of drains

The plane layout of the drains can be arranged in equilateral triangle or square, as shown in Figure 7-4. The relationship between the diameter of the effective drainage cylinder of a sand well d_e and the drainage spacing l shall be taken as follows:

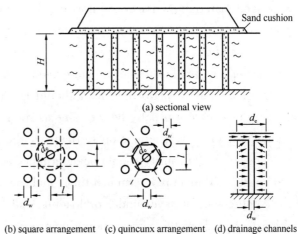

Figure 7-4 Layout of drains

| Quincunx arrangement, | $d_e = 1.05l$ | (7-8) |
| Square arrangement, | $d_e = 1.13l$ | (7-9) |

Because the quincunx arrangement is more compact and effective than the square arrangement, the quincunx arrangement is widely used. The layout area of the sand wells should be slightly larger than that of the building foundation, and the expanded range can be increased by 2-4 m from the foundation contour.

(4) Sand drain depth

The depth of the sand drains should be determined according to the stability and deformation requirements of the building on the foundation and the construction period. For the project controlled by the anti-sliding stability of foundation, the depth of sand drain should at least exceed the most dangerous sliding surface by 2 m. For buildings controlled by deformation, if the thickness of compressed soil layer is not large, sand drains should pass through the compressed soil layer. For deep compressed soil layer, the depth of sand drains should be determined according to the deformation eliminated within the limited preloading time. If the construction equipment conditions cannot reach the design depth, methods such as overload preloading can be used to meet the engineering requirements. If the thickness of soft soil layer is not large or the soft soil layer contains more thin silt interlayer, the vertical drainage body may not be set when the consolidation rate is expected to meet the requirements of the construction period.

(5) Sand cushion

A drainage sand cushion should be laid on the top surface of thesand wells. It is used to connect the sand wells to form a smooth drainage surface and discharge the water to the outside of the site. The thickness of sand cushion should not be less than 0.5 m. During underwater construction, the thickness of sand cushion is generally about 1.0 m. In order

Chapter 7 Ground Improvement

to save sand materials, vertical and horizontal sand ditches connected with sand wells can also be used to replace the whole sand cushion. The height of sand ditches is generally 0.5-1.0 m, and the width of sand ditches is twice the diameter of sand wells.

2. Determine the loading level, step-loading and loading rate

The magnitude of preloading load should be determined according to the design requirements, which can usually be the same as the basement pressure of the building. For buildings with strict restrictions on settlement, the overload preloading method should be used. The overload quantity should be determined according to the deformation required to be eliminated within a predetermined time, and the effective vertical pressure of each point of the compressed soil layer under preloading load should be equal to or greater than the additional pressure of the corresponding point caused by the building load.

The loading area should not be less than the area surrounded by the outer edge of the building foundation, so as to ensure that the foundation within the building is uniformly reinforced.

The loading rate should be adapted to the increasing strength of foundation soil, and the nextstep load should be applied after the foundation reaches a certain degree of consolidation under the previous load.

3. Calculate the consolidation degree, strength growth and deformation of foundation

(1) Calculation of consolidation degree of foundation

The average consolidation degree of the foundation corresponding to the total load at a certain time t can be calculated according to the following equation in the case of one-step or multi-step constant loading rate:

$$\overline{U}_t = \sum_{i=1}^{n} \frac{\dot{q}_i}{\Sigma \Delta p} \left[(T_i - T_{i-1}) - \frac{\alpha}{\beta} e^{-\beta \cdot t} (e^{\beta T_i} - e^{\beta T_{i-1}}) \right] \tag{7-10}$$

where

$\overline{U}_t$ ——the average consolidation degree of the foundation at time t;

$\dot{q}_i$ ——loading rate of the ith step load, kPa/d;

$\Sigma \Delta p$ ——total load levels at all steps, kPa;

T_{i-1}, T_i ——the starting and ending time of the ith step load (calculated from zero) respectively (d); T_i should be changed to t when calculating the consolidation degree at a certain time in the process of the ith step load;

α, β ——parameters, according to the soil drainage consolidation conditions, which are calculated according to Table 7-4.

(2) Calculation of foundation strength and deformation

Under preloading load, the shear strength of a point τ_{ft} in the normally consolidated saturated cohesive soil foundation at any time can be calculated as follows:

$$\tau_{ft} = \tau_{f0} + \Delta\sigma_z \cdot U_t \tan\varphi_{cu} \tag{7-11}$$

where

τ_{f0} ——natural shear strength of foundation soil, kPa;

$\Delta\sigma_z$ ——additional vertical stress at this point caused by preloading load, kPa;

U_t ——consolidation degree of the soil at this point;

φ_{cu} ——internal friction angle of soil obtained by triaxial consolidation undrained test, °.

Parameters of α and β   Table 7-4

Parameters \ Drainage Condition	Vertical drainage consolidation $\overline{U}_t > 30\%$	Radial drainage consolidation	Vertical and radial drainage consolidation (sand drains pass through the compressed soil layer)	Note
α	$\dfrac{8}{\pi^2}$	1	$\dfrac{8}{\pi^2}$	$F_n = \dfrac{n^2}{n^2-1}\ln(n) - \dfrac{3n^2-1}{4n^2}$ C_V——vertical drainage consolidation coefficient of soil, cm/s; C_h——horizontal drainage consolidation coefficient of soil, cm/s; H——vertical drainage distance of soil layer, cm; n—— drain spacing ratio
β	$\dfrac{\pi^2 C_v}{4H^2}$	$\dfrac{8C_h}{F_n d_e^2}$	$\dfrac{8C_h}{F_n d_e^2} + \dfrac{\pi^2 C_v}{4H^2}$	

(3) The final settlement of foundation under preloading load can be calculated as follow:

$$s_f = \xi \sum_{i=1}^{n} \frac{e_{0i} - e_{1i}}{1 + e_{0i}} h_i \tag{7-12}$$

where

s_f ——final vertical deformation, m;

e_{0i} ——the void ratio corresponding to the self-weight stress of ith soil layer;

e_{1i} ——the void ratio corresponding to the sum of the self-weight stress and additional stress of the ith soil layer;

h_i ——thickness of the ith soil layer, m;

ξ ——empirical coefficient, the value of $\xi = 1.1$-1.4 can be taken for normally consolidated saturated cohesive soil foundation. When the load is large and the soil is weak, take a larger value, otherwise take a smaller value.

7.3.3 Design of vacuum preloading method

The design of vacuum preloading method includes the following contents.

(1) Vertical drainage body. Vertical drainage body must be set when vacuum preload-

ing method is used. The vertical drainage body can adopt bagged sand drains with a diameter of 700 mm, or ordinary sand drains or prefabricated strip drains. Its spacing can be selected according to the spacing of sand drain or prefabricated strip drains in surcharge preloading method. The depth of sand drains shall be determined according to the settlement completed during preloading and the requirements of foundation stability of the proposed building. Sand in the sand drains should be medium and coarse sand. Its permeability coefficient k should be greater than 1×10^{-2} cm/s.

(2) Preloading area. The edge of vacuum preloading area should be larger than the contour line of building foundation. The increase on each side should not be less than 3.0 m. The preloading area of each block should be as large and square as possible.

(3) The vacuum degree in the membrane of vacuum preloading should be kept above 650 mmHg, which is equivalent to the vacuum pressure of 86.7 kPa. The vacuum pressure should be evenly distributed. The average consolidation degree of soil layer within the depth of sand drains should be greater than 90%.

(4) Settlement calculation. Firstly, the settlement of natural foundation under building load is calculated, and then the settlement completed during vacuum preloading is calculated. The difference between them is the possible settlement under building service load after preloading.

Example 7-2:

There is a saturated clay layer with a thickness of 10 m, and the bottom of the soil layer is an impermeable layer. It is planned to adopt sand drain. The diameter of the sand drain $d_w = 20$ cm, the length $l = 10$ m, and the effective drainage diameter $d_e = 200$ cm. The consolidation coefficient of soil layer is $c_v = 8.34 \times 10^{-4}$ cm²/s and $c_h = 8.34 \times 10^{-3}$ cm²/s. The total preloading load is $p = 100$ kPa, and it is loaded in two stages at the same speed. The first step load $p_1 = 60$ kPa, and the loading is completed in 10 days. The second-step load is intended to be applied when the consolidation degree of the reinforced soil reaches 80% under the first-step load. Please calculate the time when the second-step load is applied (don't need to consider well resistance and smearing).

Solution:

According to the calculation equation of average consolidation degree,

$$\overline{U}_t = \sum_{i=1}^{n} \frac{\dot{q}_i}{\Sigma \Delta p} \left[(T_i - T_{i-1}) - \frac{\alpha}{\beta} e^{-\beta \cdot t} (e^{\beta T_i} - e^{\beta T_{i-1}}) \right]$$

Loading rate of the first-step load: $\dot{q}_1 = 60/10 = 6$ kPa/d

Total value of load: $\Sigma \Delta p = 60$ kPa. $T_0 = 0$, $T_1 = 10$

Drain spacing ratio: $n = d_e/d_w = 200/20 = 10$

α and β calculated based on Table 7-4:

$$F_n = \frac{n^2}{n^2 - 1} \ln(n) - \frac{3n^2 - 1}{4n^2} = \frac{10^2}{10^2 - 1} \times \ln(10) - \frac{3 \times 10^2 - 1}{4 \times 10^2} = 1.578$$

$$\alpha = 8/\pi^2 = 0.81$$

$$\beta = \frac{8C_h}{F_n d_e^2} + \frac{\pi^2 C_v}{4H^2} = \frac{8 \times 8.34 \times 10^{-3} \times 8.64}{1.578 \times 2^2} + \frac{\pi^2 \times 8.34 \times 10^{-4} \times 8.64}{4 \times 10^2} \approx 0.0915 \text{ d}^{-1}$$

The average consolidation degree of foundation soil is:

$$\overline{U}_t = 0.8 = \frac{6}{60}\left[(10-0) - \frac{0.81}{0.0915}e^{-0.0915 \times t}(e^{0.0915 \times 10} - e^0)\right]$$

$$\Rightarrow t = 21\text{d}$$

7.4 Dynamic compaction

Dynamic compaction (DC) uses the kinetic energy of a free falling heavy weight, typically 10-40t, dropped from heights ranging from 15 to 40 m, to rearrange the coarse soil particles thereby decreasing the soil void ratio and increasing stiffness. This method, developed by Louis Menard in the late 1960s, is applicable to granular soils below the groundwater to depths of up to 15 m, and can also be applied to non-saturated, fined grained soils such as silts and clays.

Dynamic compaction replacement method is different from dynamic compaction, which backfills coarse-grained materials such as block stone, crushed stone and steel slag in tamping pit, and tamping to form continuous dynamic compaction replacement pier.

Dynamic compaction and dynamic compaction replacement method have the advantages of simple construction, good strengthening effect, and economic use. They are widely used in various soil treatment around the world.

7.4.1 Strengthening mechanism of dynamic compaction

For dynamic compaction, there are three different strengthening mechanisms: dynamic compaction, dynamic consolidation, and dynamic replacement, which depend on the soil type and the construction technology of dynamic compaction.

1. Dynamic compaction

The use of dynamic compaction to strength porous, coarse-grained, and unsaturated soil is based on the mechanism of dynamic compaction. It uses dynamic impact loads to reduce the pores in the soil and make the soil dense, thereby improving the strength of the foundation soil. The compaction process of unsaturated soil is the process in which the gas (air) in the soil is squeezed out, and the compaction deformation is mainly caused by the relative displacement of soil particles.

2. Dynamic consolidation

When using the dynamic compaction to improve fine-grained saturated soil, it relies on the theory of dynamic consolidation. During dynamic compaction, the huge impact energy generates large stress waves in the soil and damages the original structure of the soil, which cause local liquefaction and cracking of the soil, thereby increasing drainage channels and accelerating the discharge of pore water. As the excess pore water pressure dissipates, the soil gradually consolidates. The strength of soft soil is improved due to its thixotropy.

3. Dynamic replacement

Dynamic replacement is the use of the impact force generated during compaction to forcibly fill sand, gravel, and other materials into saturated soft soil, thus forming a pile column or dense stone layer. At the same time, the saturated soft soil in the underlying layer that has not been replaced is drained and consolidated under the action of force. It becomes denser. Thus, the bearing capacity of the foundation is increased and the settlement is reduced.

7.4.2 Design of dynamic compaction

The design of dynamic compaction is given in the following.

1. Effective compact depth

The effective compact depth is not only an important basis for selecting ground improvement methods, but also an important parameter reflecting the improvement effect. The effective compact depth of dynamic compaction method can be estimated according to the modified Menard equation:

$$H = \alpha \sqrt{Mh} \qquad (7\text{-}13)$$

where

H——effective compact depth, m;

M——hammer weight, kN;

h——falling distance, m;

α——correction coefficient, which is related to the properties of the foundation soil;

It can be 0.5 for soft soil and 0.34-0.5 for loess.

In fact, there are many factors that affect the effective compact depth of dynamic compaction. Besides the hammer weight and falling distance, there are also the properties of foundation soil, the thickness and layer order of different soil, groundwater level and other design parameters of dynamic compaction. Therefore, the effective compact depth of dynamic compaction should be determined according to field trial compaction or local experience. In the absence of test data or experience, it can also be estimated according to Table 7-5.

Effective compact depth of dynamic compaction Table 7-5

Single tamping energy $E/\text{kN} \cdot \text{m}$	Coarse grained soil such as gravel soil and sandy soil	Fine grained soil such as silt, cohesive soil and collapsible loess
1000	4.0-5.0	3.0-4.0
2000	5.0-6.0	4.0-5.0

7.4 Dynamic compaction

continued

Single tamping energy $E/\text{kN} \cdot \text{m}$	Coarse grained soil such as gravel soil and sandy soil	Fine grained soil such as silt, cohesive soil and collapsible loess
3000	6.0-7.0	5.0-6.0
4000	7.0-8.0	6.0-7.0
5000	8.0-8.5	7.0-7.5
6000	8.5-9.0	7.5-8.0
8000	9.0-9.5	8.0-8.5
10000	9.5-10.0	8.5-9.0
12000	10.0-11.0	9.0-10.0

Note: The effective compact depth of dynamic compaction should be calculated from the initial tamping surface. When the single tamping energy E is greater than 12000 kN · m, the effective compact depth should be determined through experiments.

2. Hammer and drop distance

Single tamping energy is the product of hammer weight M and falling distance h. Generally speaking, when tamping, it is best to have a large hammer weight and drop distance. When the single tamping energy is larger and the tamping times are correspondingly reduced, the strengthening effect and technical economy are better. The total tamping energy of the whole site (i. e. hammer weight × falling distance × total tamping number) divided by the improvement area is called unit tamping energy. The unit tamping energy of dynamic compaction and dynamic compaction replacement should be comprehensively considered according to the type of foundation soil, structure type, load level and required compaction depth. It can be determined through experiments.

3. Arrangement and spacing of tamping spots

The tamping spots of dynamic compaction can be arranged in equilateral triangle, isosceles triangle or square according to the plane shape of the site. At the same time, the layout of tamping spots should consider the walking passage during construction. The equilateral triangle or square should be adopted in the layout of dynamic replacement pier. The pad foundation or strip foundation can be arranged according to the shape and width of the foundation.

The area of dynamic compaction and dynamic compaction replacement should be larger than that of building foundation. For general buildings, the width of each side beyond the outer edge of the foundation should be 1/2-2/3 of the compact depth, and it should not be less than 3 m. For liquefiable subgrade, the compaction width outside of the foundation edge should not be less than 5 m.

The spacing of tamping spots in the first time of dynamic compaction can be 2.5-3.5 times of the hammer diameter. In the second time, the tamping spots should be located between the tamping spots in the first time. And the spacing of tamping spots in subsequent times can be appropriately reduced. For projects with deep compact depth or large single tamping energy, the spacing between tamping spots in the first time should be appropriately increased. The spacing of tamping spots is generally based on the properties of foundation soil and the depth required to be compacted.

The distance between the dynamic replacement piers should be selected according to the load and the bearing capacity of the natural soil. It can be 2-3 times of the hammer diameter when it is fully arranged. For pad foundation or strip foundation, the distance between the dynamic replacement piers can be 1.5-2.0 times of the hammer diameter.

4. Tamping number and tamping times

The tamping number at a single spot should be determined according to the relationship curve of tamping number and tamping settlement obtained from field trial experiments. The following requirements should be met.

(1) The average settlement of the last two blows should meet the requirements given in Table 7-6. When single tamping energy is greater than 12000 kN · m, it should be determined through experiments.

The average settlement of the last two blows   Table 7-6

Single tamping energy $E/kN \cdot m$	The average settlement of the last two blows should not greater than/mm
$E < 4000$	50
$4000 \leqslant E < 6000$	100
$6000 \leqslant E < 8000$	150
$8000 \leqslant E < 12000$	200

(2) The ground around the tamping spot should not be excessively uplifted.

(3) It is not difficult to raise the hammer because the tamping pit is too deep.

The tamping number of dynamic compaction replacement method should be determined through field trial tamping experiments. The following requirements shall be met: (a) The pier bottom penetrates the soft soil layer and reaches the designed pier length; (b) The accumulated tamping settlement is 1.5-2.0 times of the designed pier length; (c) The average settlement of the last two blows is not greater than the specified value of dynamic compaction.

The tamping times should be determined according to the properties of foundation soil. Generally, the tamping times of 2-4 can be used. For fine-grained soil with poor permeability, the tamping times can be increased appropriately if necessary. Finally, it is tamped twice with low energy.

5. Cushion layout

When the topsoil on the site is weak or the groundwater level is high, it is required that the site must have a slightly hard surface layer before dynamic compaction. It is used to support the lifting equipment, facilitate the diffusion of the tamping energy, and increase the distance between the groundwater level and the ground surface. Therefore, it is necessary to lay a cushion sometimes. For the gravel soil layer with the ground water level 2 m below, dynamic compaction can be directly implemented without laying cushion. For saturated cohesive soil with high groundwater level and saturated sandy soil which is easy to liquefy and flow, it is necessary to lay sand, gravel or gravel cushion for dynamic compaction, otherwise the soil will flow. The thickness of cushion depends on the soil conditions of the site, the weight and shape of hammer, and etc. The thickness of cushion can also be reduced if the soil conditions are good, the hammer is small or the suction force is small when lifting. The cushion thickness is generally 0.5-2.0 m.

6. Intermittent time

For projects that need to betamped twice or more times, there should be a certain time interval between the two tamping. The interval time between tamping times depends on the time required for the pore water pressure in the strengthening soil layer to dissipate. For sandy soil, the peak pore water pressure appears at the moment after tamping, and the dissipation time is only 2-4 min. Therefore, for sandy soil with large permeability, the intermittent time between tamping times is very short and it can be tamped continuously.

For cohesive soil, because the dissipation of pore water pressure is slow, the pore water pressure will be superimposed when the tamping energy increases gradually. Its intermittent time depends on the dissipation of pore water pressure, which is generally 2-3 weeks. At present, some projects use bagged sand drains (or prefabricated strip drains) on the cohesive soil site in order to accelerate the dissipation of pore water pressure and shorten the intermittent time.

7.5 Composite foundation

7.5.1 Concept and classification of composite foundation

The concept of composite foundation

Composite foundation refers to an artificial foundation in which part of the soil is reinforced or replaced to form a reinforcing body, and the reinforcing body and surrounding soil jointly bear the upper load and have compatible deformation. The composite foundation has two basic characteristics: (a) The reinforcement area is composed of the reinforcing body and its surrounding soil, which is heterogeneous and anisotropic; (b) The reinforcing body and its surrounding soil bear the load together and have compatible deformation. The first one characteristic of the composite foundation makes it different from natural foundation, and the second one makes it different from pile foundation. By comparing natural foundation, composite foundation and pile foundation, there are essential differences in their mechanical characteristics, as shown in Figure 7-5. In composite foundation, the reinforcing body of pile and foundation are often not directly connected, and they are separated by gravel or gravel cushion. However, in the pile foundation, the pile body and the foundation (cap) are directly connected to form a whole. The main bearing layer of composite foundation is in the reinforcement range, yet the main bearing layer of pile foundation is in a certain range below the pile tip. Because the most basic assumption of composite foundation is the compatiable deformation of piles and soil around piles, there is no pile group effect in composite foundation similar to that in pile foundation.

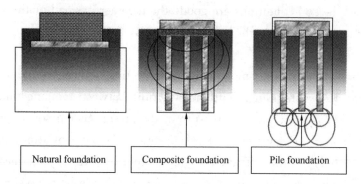

Figure 7-5 Comparison of natural foundation, composite foundation and pile foundation

According to the reinforcement direction, composite foundation can be divided into horizontal reinforced composite foundation and vertical reinforced composite foundation, as shown in Figure 7-6. Horizontal reinforced composite foundation are formed by various reinforced materials, such as geopolymer and metal grid. Vertical reinforced composite foundation is usually formed by various pile bodies, which is usually called pile composite foundation.

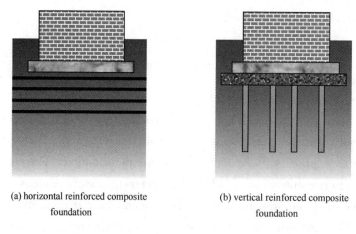

(a) horizontal reinforced composite foundation

(b) vertical reinforced composite foundation

Figure 7-6 Types of composite foundation

In the pile composite foundation, piles are the main role. There are various types of piles in foundation treatment, and their performance changes greatly. Therefore, it is appropriate to classify the composite foundation according to the types of piles. Piles can be classified according to their materials and the strength (or stiffness) of the pile body after pile formation.

According to the type of material, piles can be divided into (a) discrete material piles-such as gravel piles and sand piles; (b) cement-soil piles-such as cement-soil mixing piles and jet grouting piles; and (c) con crete piles-root piles, CFG piles, and etc.

According to the strength (or stiffness) of the pile, piles can be divided into (a) flexible piles-granular soil pile belongs to this kind of pile; (b) semi-rigid piles-such as cement-soil pile; and (c) rigid piles-such as concrete pile.

The cement content in semi-rigid piles will directly affect the strength of piles. When the mixing amount is small, the characteristics of the pile are similar to those of the flexible pile. However, when the mixing amount is large, it is similar to rigid pile.

The composite foundation composed of flexible piles and soil between piles can be called flexible pile composite foundation, and the other two are semi-rigid pile composite foundation and rigid pile composite foundation respectively.

7.5.2 Reinforcement mechanism of composite foundation

No matter what kind of composite foundation, it has one or more of the following functions, which are given as follows.

1. Pile function

Because the stiffness of the pile in the composite foundation is greater than that of the surrounding soil, the stress in the foundation will be distributed according to the material modulus when the deformation is equal under the rigid foundation. Therefore, the pile body produces stress concentration phenomenon, most of the load is borne by the pile body, and the stress on the soil between piles is reduced accordingly. In this way, the bearing capacity of composite foundation is improved compared with that of natural foundation, and the settlement is reduced. with the increase of pile stiffness, its pile function becomes more obvious.

2. Cushion function

The structural difference between the composite foundation and the pile foundation is that the pile groups in the pile foundation are connected with the cap, while the piles in the composite foundation are transited with the shallow foundation through the cushion. The cushion of composite foundation can adjust the relative deformation of pile and soil, avoid the stress concentration of pile caused by load, and effectively ensure the normal work of pile. In the foundation where the pile does not pass through the whole soft soil layer, the effect of cushion is particularly obvious.

3. Accelerated consolidation

In addition to gravel piles and sand piles, which have good water permeability and can accelerate the foundation consolidation, cement-soil piles and concrete piles can also accelerate the foundation consolidation to some extent. The reason is that foundation consolidation is not only related to the drainage performance of foundation soil, but also related to the deformation characteristics of foundation soil. It can be clearly seen by the formula of consolidation coefficient c_v [$c_v = k(1+e_0)/\gamma_w \cdot a$]. Although the cement-soil pile will reduce the permeability coefficient k of foundation soil, it will also reduce the compressibility coefficient a of foundation soil. Generally, the latter will reduce more than the former. Therefore, the consolidation coefficient c_v of cement-soil after reinforcement is greater than that of in-situ soil before reinforcement, which can also accelerate consolidation.

4. Compaction

The piles, such as sand piles, soil piles, lime piles, and gravel piles, can make the soil between piles play a certain compaction role due to the vibration and extrusion during construction. In addition, quicklime and cement powder in lime piles and jet mixing piles can absorb water, release heat and expand, which also have a certain compaction effect on soil between piles.

5. Reinforcement

All kinds of pile-soil composite foundation can not only improve the bearing capacity of foundation, but also improve the shear strength of soil and the anti-sliding ability of soil slope. At present, deep mixing piles, powder jet mixing piles and sand piles are widely used in the reinforcement of subgrade or embankment, which make use of the reinforcement effect of piles in composite foundation.

7.5.3 Design parameter of composite foundation

The design parameters of composite foundation mainly include area replacement ratio and pile-soil stress ratio.

1. Area replacement ratio m

In the composite foundation, the representative unit can be taken as one pile and the soil reinforced by the pile. The area replacement ratio is defined as ratio of the cross-sectional area (A_p) of the pile body to the composite foundation area (A_e) reinforced by the pile. That is:

$$m = \frac{A_p}{A_e} = \frac{d^2}{d_e^2} \qquad (7\text{-}14)$$

where

d ——average diameter of pile body, m;

d_e ——equivalent circle diameter of the foundation area reinforced by one pile, m.

Usually, within the reinforcement area, the piles are arranged in equilateral triangles or squares, as shown in Figures 7-7(a) and (b). According to the layout of the pile, The d_e can be calculated as follows:

Equilateral triangular arrangement: $d_e = 1.05 s_a$ (7-15)

Square arrangement: $d_e = 1.13 s_a$ (7-16)

where

s_a ——pile spacing, m.

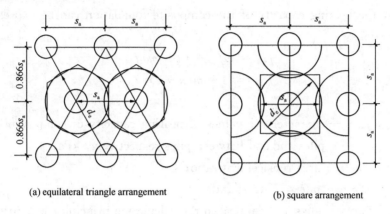

(a) equilateral triangle arrangement (b) square arrangement

Figure 7-7 Pile layout in composite foundation

2. Pile-soil stress ratio n

The pile-soil stress ratio n refers to the ratio of the average vertical stress of piles (σ_p) in composite foundation to the average vertical stress of soil between piles (σ_s). The pile-soil stress ratio is an important design parameter in composite foundation, which is related to the calculation of bearing capacity and deformation. The factors affecting the pile-soil stress ratio include load level, pile-soil modulus ratio, area replacement ratio, natural soil strength, pile length, consolidation time and cushion. Assuming that the vertical strain of the pile and the soil between piles is equal under the rigid foundation, the calculation equation of the pile-soil stress ratio n can be obtained as follows:

$$n = \frac{\sigma_p}{\sigma_s} = \frac{E_p}{E_s} \tag{7-17}$$

where

E_p, E_s ——the compressive modulus of the pile and the soil between piles respectively, MPa.

7.5.4 Bearing capacity of composite foundation

The bearing capacity of composite foundation should generally be determined by on-site composite foundation load test, which should be carried out according to the *Technical Code for Ground Treatment of Buildings* (*JGJ* 79—2012). The preliminary design can also be estimated by composite summation method. The composite summation method firstly determines the bearing capacity of piles and the bearing capacity of soil between piles respectively, and then superposes these two parts to obtain the bearing capacity of composite foundation. On the basis of this theory, there are two calculation methods for different types of composite foundation. One is stress composite method, and the other is deformation composite method.

The stress composite method assumes that when the composite foundation reaches its bearing capacity, the piles and the soil between piles also reach their respective bearing capacity. Thus, the bearing capacity of the composite foundation can be expressed by the following equations:

$$f_{spk} = mf_{pk} + (1-m)f_{sk} \qquad (7\text{-}18)$$

$$f_{spk} = [1 + m(n-1)]f_{sk} \qquad (7\text{-}19)$$

where

f_{spk}, f_{pk}, f_{sk} —— characteristic values of bearing capacity of composite foundation, piles and soil between piles respectively, kPa;

m —— area replacement ratio;

n —— pile-soil stress ratio.

The above two formulas are identical in the calculation principle, but different in application. The application of Equation 7-18 needs to obtain the characteristic value of bearing capacity of single pile through calculation or test, while the application of Equation 7-19 needs to obtain the pile-soil stress ratio.

The deformation composite method assumes that when the composite foundation reaches its bearing capacity, the bearing capacity of the pile is fully exerted while the bearing capacity of the soil is not fully exerted. It means the pile and the soil between piles do not reach their respective bearing capacity at the same time. Because the stress of soil is related to deformation, the formula for solving the bearing capacity of composite foundation by strain composite method is:

$$f_{spk} = \lambda m \frac{R_a}{A_p} + \beta(1-m)f_{sk} \qquad (7\text{-}20)$$

where

λ —— bearing capacity mobilization coefficient of a single pile, which can be adopted according to regional experience;

β —— the bearing capacity reduction factor considering that the deformation of soil between piles is less than 1.0;

R_a —— characteristic value of vertical bearing capacity of single pile, kN;

A_p —— cross-sectional area of single pile, m².

The vertical bearing capacity of single pile can be calculated using the following equation:

$$R_a = u_p \sum_{i=1}^{n} q_{si} l_{pi} + \alpha_p q_p A_p \qquad (7\text{-}21)$$

where

u_p ——perimeter of the single pile, m;

n ——the number of soil layers passed within pile length;

q_{si} ——sleeve resistance of the ith layer soil around the pile, kPa;

l_{pi} ——thickness of the ith layer soil within pile length, m;

q_p ——tip resistance of the foundation soil at the pile tip, kPa;

α_p ——the mobilization coefficient of tip resistance, which should be determined according to the regional experience.

The pile strength of composite foundation with cohesive strength piles should meet the requirements of Equation 7-22 for verifying pile strength. When the bearing capacity of composite foundation is corrected in depth, the strength of reinforced pile body should meet the requirements of Equation 7-23.

$$f_{cu} \geqslant 4\frac{\lambda R_a}{A_p} \tag{7-22}$$

$$f_{cu} \geqslant 4\frac{\lambda R_a}{A_p}\left(1+\frac{\gamma_m(d-0.5)}{f_{spa}}\right) \tag{7-23}$$

where

f_{cu} ——average cubic compressive strength of standard cubic pile test block with its side length of 150 mm, kPa;

γ_m ——weighted average unit weight of soil above the foundation bottom, kN/m^3, and the buoyant unit weight is taken below the groundwater level;

d ——buried depth of foundation, m;

f_{spa} ——bearing capacity characteristic value of composite foundation after depth correction, kPa.

According to *Technical Code for Ground Treatment of Builidings* (JGJ 79—2012), the calculation method selection of bearing capacity of various composite foundations is specified in detail. The stress composite method is adopted for flexible pile composite foundations such as sand-gravel pile composite foundation and lime pile composite foundation. The deformation composite method is used for semi-rigid and rigid pile composite foundations such as cement-soil mixing pile, CFG pile composite foundation and jet grouting pile composite foundation.

When calculating the bearing capacity of composite foundation, it is also necessary to determine the bearing capacity of pile f_{pk} and the bearing capacity of soil between piles f_{sk}. For the bearing capacity of pile f_{pk}, it can be measured by field pile load test. And it can be estimated by the equation of bearing capacity of single pile. For different types of piles, the calculation equations are different, which are related to specific composite foundation types. It should be pointed out that the bearing capacity of soil between piles f_{sk} is the bearing capacity of soil in the improved foundation bearing layer. When the bearing capacity of soil between piles changes little before and after improvement, the uncorrected bearing capacity can be directly used. However, if the pile body has obvious extrusion effect, the bearing capacity of soil between piles will be changed greatly after improvement. It is necessary to determine the bearing capacity of soil after improvement.

For the composite foundation, the bearing capacity characteristic value of the composite foundation is corrected only according to the buried depth of the foundation, as shown in the following equation:

$$f_{spa} = f_{spk} + \eta_d \gamma_m (d - 0.5) \tag{7-24}$$

where

η_d ——depth correction coefficient of foundation bearing capacity. It can be taken as 1.0 for composite foundations.

7.5.5 Settlement calculation of composite foundation

The settlement calculation of composite foundation is the same as that of general multi-layer foundation, and the layering of composite soil layer is the same as that of natural foundation. According to *Technical Code for Ground Treatment of Builidings* (*JGJ 79—2012*), the calculated depth for composite foundation settlement should be greater than that of composite soil layer. If there is still soft soil layer under the determined calculated depth, the calculation should be continued. The final settlement of composite foundation can be calculated according to Equation 7-25:

$$s = \psi_{sp} s' \tag{7-25}$$

where

s ——final settlement of composite foundation;

ψ_{sp} ——the settlement empirical coefficient of composite foundation. It can be determined according to the regional experience. When there is no regional experience, it can be taken according to Table 7-7 according to the equivalent compression modulus within the calculated depth $\overline{E_s}$;

s' ——the deformation of composite foundation calculated by using layered summation method.

Empirical coefficient for settlement calculation of composite foundation ψ_{sp} Table 7-7

$\overline{E_s}$ /MPa	4.0	7.0	15.0	20.0	35.0
ψ_{sp}	1.0	0.7	0.4	0.25	0.2

Compressive modulus of composite soil layer E_{sp} can be calculated according to the following equations:

$$E_{sp} = \zeta \cdot E_s \tag{7-26}$$

$$\zeta = \frac{f_{spk}}{f_{ak}} \tag{7-27}$$

where

E_{sp} ——compressive modulus of composite soil layer, MPa;

E_s ——compressive modulus of natural soil, MPa;

f_{ak} ——bearing capacity of natural soil directly below the foundation, kPa.

The equivalent compression modulus within the calculated depth for composite foundation $\overline{E_s}$ is calculated according to the following equation:

$$\overline{E}_s = \frac{\sum_{i=1}^{n} A_i + \sum_{j=1}^{m} A_j}{\sum_{i=1}^{n} \dfrac{A_i}{E_{spi}} + \sum_{j=1}^{m} \dfrac{A_j}{E_{si}}} \tag{7-28}$$

where

A_i ——the integral value of the additional stress coefficient of the ith-layer soil along the thickness of the soil layer within the scope of the reinforced soil layer;

A_j ——the integral value of the additional stress coefficient of the jth layer of soil along the thickness of the soil layer below the reinforced soil layer;

m ——the number of layers below the reinforced soil layer within the calculated depth for composite foundation settlement;

n ——the number of layers of reinforced soil layer within the calculated depth for composite foundation settlement.

7.5.6 Sand-gravel pile composite foundation

In sand-gravel pile composite foundation, crushed stone, sand or mixture of sand and gravel are extruded into the formed holes to form a composite foundation. The dense sand and gravel columns are the vertical reinforcement of composite foundation. According to *Technical Code for Ground Treatment of Builidings* (*JGJ 79—2012*), sand-gravel pile composite foundation is suitable for compaction of loose sand, silt, cohesive soil, plain fill, miscellaneous fill and other foundations. For projects with low deformation control on saturated clay foundation, sand-gravel pile composite foundation can also be used. Sand pile composite foundation can also be used to treat liquefiable foundation.

The reinforcing mechanism of sand-gravel piles is dependent on the pile-forming methods and soil type. The function of sand-gravel piles in strengthening foundation can be reflected in the following aspects: (1) compaction; (2) vibration compaction; (3) transformation; (4) drainage.

The design of sand-gravel pile composite foundation includes the following contents.

1. Scope of reinforcement

The scope of reinforcement should be determined according to the importance of the building, site conditions and type of foundation, which are usually larger than the outer edge of foundation. For general foundation, the reinforcement area should be greater than the foundation bottom area, and the reinforcement width should be increased by 1-3 rows

Chapter 7 Ground Improvement

of piles outside of foundation. For liquefiable foundation, the widening width at the outer edge of foundation should not be less than 1/2 of the thickness of liquefiable soil layer and not less than 5 m.

2. Pile diameter and arrangement

The diameter of sand-gravel piles should be determined according to the soil conditions of foundation and pile-forming equipment. It can be calculated according to the amount of filling material used in each pile. The diameter of sand-gravel pile can be 300-1000 mm, and the larger diameter should be adopted for saturated cohesive soil foundation.

For large area full treatment, the piles should be arranged in equilateral triangle. For pad or strip foundation, the piles should be arranged in square, rectangle or isosceles triangle. For circular or annular foundations (such as oil tank foundations), radial layout should be used.

3. Spacing between sand-gravel piles

Because the reinforcing mechanism of sand-gravel piles in loose sand and soft cohesive soil is different, the calculation method of pile spacing is also different. In sandy soil foundation, the basic assumption is that the soil particles increase and the volume remains unchanged after compaction, so as to control the void ratio after improvement. The pile spacing can be calculated according to the required void ratio.

When arranged in an equilateral triangle,

$$s = 0.95 \xi d \sqrt{\frac{1+e_0}{e_0 - e_1}} \tag{7-29}$$

When arranged in a square,

$$s = 0.89 \xi d \sqrt{\frac{1+e_0}{e_0 - e_1}} \tag{7-30}$$

$$e_1 = e_{max} - D_{rl}(e_{max} - e_{min}) \tag{7-31}$$

where

s ——spacing between gravel piles, m;

D ——diameter of gravel pile, m;

ξ ——correction coefficient, which is 1.1-1.2 when considering vibration subsidence and compaction; 1.0 can be taken when the vibration sinking compaction effect is not considered;

e_0 ——porosity ratio of sand before foundation treatment: It can be determined by undisturbed soil test or by dynamic or static sounding test;

e_1 ——the required void ratio after foundation treatment;

e_{max}, e_{min} ——the maximum and minimum void ratio of sand;

D_{rl} ——the required relative density of sand. Generally 0.70-0.85 can be taken.

In cohesive soil foundation, the pile spacing can be calculated according to the requirement of area replacement ratio.

When arranged in an equilateral triangle,
$$s = 1.08\sqrt{A_e} \tag{7-32}$$
When arranged in a square,
$$s = \sqrt{A_e} \tag{7-33}$$
where

A_e——the treatment area undertaken by one pile, m².
$$A_e = A_p/m \tag{7-34}$$
where

A_P——the cross-sectional area of sand-gravel pile, m²;

m——area replacement ratio.

4. Length of sand-gravel pile

The length of sand-gravel pile can be determined according to the engineering requirements and engineering geological conditions. Generally, it should not be less than 4 m. When the thickness of soft soil layer is not large, the length of gravel pile should pass through soft soil layer. When the thickness of soft soil layer is large, the length of gravel pile should meet the requirements that the settlement of the treated foundation should not exceed the allowable settlement of the building and the bearing capacity of the soft underlying layer can be meet. Meanwhile, the reinforcement depth should not be less than 2 m below the most dangerous sliding surface for the projects controlled by stability.

5. Material

The materials of the pile body can be made from local materials. Hard materials such as gravel, pebble, breccia, round gravel, gravel sand, coarse sand, medium sand or stone chips can be used. The silt content shall not be greater than 5% and the maximum particle size should not be greater than 50 mm.

6. Cushion

After the construction of sand-gravel piles is completed, a sand-gravel cushion with a thickness of 300-500 mm shall be laid on the bottom of the foundation. It should be laid in layers and vibrated with a flat vibrator. Temporary cushion for construction should be laid on the soft soil layer to guarantee the normal running and operation of construction machinery.

7.5.7 Cement-soil mixed pile composite foundation

In cement-soil mixing pile composite foundation, cement is used as the main curing agent, and the curing agent (slurry or powder) and foundation soil are forcibly mixed by a special deep mixing machine. Thus, soft soil is hardened into a pile with integrity, water stability and high strength. It is suitable for treating normal consolidated silt and silt soil, silt, saturated loess, plain fill, cohesive soil and saturated loose sandy soil foundation without flowing groundwateris. According to the type of mixing methods, cement-soil mixed pile can be divided into deep mixing pile (wet method for short) and powder spray mixing pile (dry method for short). Dry method is not suitable when the natural moisture content of foundation soil is less than 30% (loess moisture content is less than 25%) or more than 70% or the pH value of groundwater is less than 4. Cement-soil mixed column can be used as vertical reinforcement composite foundation, such as retaining wall for excavation, passive area reinforcement, impervious curtain, mass cement stabilized soil and so on.

The characteristics of cement-soil mixed pile composite foundation can be given as follows. (a) There is no vibration, no noise, no pollution to the surrounding environment, no lateral extrusion to soft soil, and little influence on adjacent buildings; (b) Various reinforcement shapes such as column, wall, grid and block can be flexibly adopted according to the requirements of the superstructure; (c) It can effectively improve the foundation strength (When the cement content is 8% and 10%, the reinforcement strength is 0.24 MPa and 0.65 MPa respectively, while the natural soft soil strength is only 0.006 MPa); (d) The construction machines and tools are relatively simple, and the construction period is short. The total cost is low, and the benefits are remarkable.

The design of cement-soil mixed pile composite foundation includes the following items.

1. Requirements for geological survey

Before determining the treatment scheme, detailed geotechnical engineering data for the area to be treated should be collected, which includes: (a) thickness and composition of the fill layer; (b) distribution and stratification of soft soil layer; (c) groundwater level and pH value; (d) moisture content, plasticity index and organic matter content of soil.

2. Selection of pile layout

According to the characteristics of the superstructure and the requirements for the bearing capacity and deformation of the foundation, the pile layout can adopt different forms such as column, wall, grid or block. Piles can only be arranged within the plane of foundation, and the number of piles underpad foundation should not be less than three. Column reinforcement can be arranged in square and equilateral triangle.

3. Determination of pile length and diameter

The length of cement-soil mixed pile should be determined according to the requirements of upper structure for bearing capacity and deformation. It should pass soft soil layer to reach soil layer with relatively high bearing capacity. In order to improve the antisliding stability, the length of the mixed pile should be more than 2 m below the dangerous sliding arc. The depth for wet method should not be greater than 20 m and the depth for dry method should not be greater than 15 m. The diameter of cement-soil mixed pile should not be less than 500 mm.

4. Bearing capacity of cement-soil mixing pile composite foundation

The bearing capacity of the cement-soil mixed pile should be determined by field load test of a single pile or multi-pile composite foundation. In the preliminary design, it can also be estimated according to Equation 7-20. By using Equation 7-20 in cement-soil mixing pile composite foundation, f_{sk} is the bearing capacity characteristic value of soil between piles (kPa), which can be taken as the characteristic value of natural foundation bearing capacity. The β is the bearing capacity reduction coefficient of soil between piles. For silt, muddy soil and plastic flow soft soil, it can be taken as 0.1-0.4. For other soils, 0.4-0.8 can be taken. The parameter of λ can be taken as 1.0 for cement-soil mixing pile composite foundation.

The characteristic value of vertical bearing capacity of single pile should be determined by field load test. In the preliminary design, it can also be estimated according to the following two equations, and the smaller value is taken.

$$R_a = u_p \sum_{i=1}^{n} q_{si} l_i + \alpha_p q_p A_p \tag{7-35}$$

$$R_a \leqslant \eta f_{cu} A_p \tag{7-36}$$

where

R_a——characteristic value of vertical bearing capacity of a single pile, kPa;

f_{cu}——compressive strength of cubic test block with side length of 70.7 mm at 90 d age, kPa;

η——pile strength reduction coefficient which can be 0.20-0.25 for dry method and 0.25 for wet method;

u_p——circumference of single pile, m;

n——number of soil layers passed within the pile length;

q_{si}——side resistance of the ith-layer soil around the pile;

l_i——thickness of the ith-layer soil within the pile length, m;

q_p——bearing capacity of foundation soil at pile tip, kPa, which can be taken as the bearing capacity of soil at the pile tip;

α_p——bearing capacity reduction coefficient of natural foundation soil at the pile tip, which can be taken as 0.4-0.6.

Example 7-3:

There is a loose sandy soil subgrade with the initial porosity ratio of $e=0.78$ before treatment. The maximum and minimum porosity ratios of the sandy soil are 0.91 and 0.58, respectively. The subgrade is improved by using sand and gravel piles, and the relative density of the compacted sandy soil subgrade is required to reach 0.85. If the pile diameter is 0.8 m and arranged in the shape of equilateral triangle, please determine the reasonable spacing between piles.

Solution:

According to Equations 7-29 and 7-31:

The porosity ratio of sandy soil after improvement:
$$e_1 = e_{max} - D_{rl}(e_{max} - e_{min}) = 0.91 - 0.85 \times (0.91 - 0.58) = 0.63$$
The spacing between piles:
$$s = 0.95\xi d\sqrt{\frac{1+e_0}{e_0 - e_1}} = 0.95 \times 1.2 \times 0.8 \times \sqrt{\frac{1+0.78}{0.78 - 0.63}} = 3.14 \text{ m}$$

Example 7-4:

A multi-storey building adopts square raft foundation, and the geological profile and soil parameters are shown in the Figure 7-8. The cement-soil mixed piles are used to treat the soil. The diameter of the cement-soil mixed pile is 550 mm with length of 10 m. These piles are evenly distributed in a square shape. Try to determine the following two questions.

(1) Assuming that the cubic compressive strength of the test pile block is 1,800 kPa, the strength reduction coefficient of the pile body is 0.3, the characteristic bearing capacity of the natural soil at the pile end is 120 kPa, and the bearing capacity reduction coefficient is 0.5, try to determine the characteristic vertical bearing capacity of a single cement-soil mixed pile.

(2) Assuming that characteristic vertical bearing capacity of a single cement-soil mixed pile is 180 kN, the characteristic bearing capacity of soil between piles is 100 kPa, and the bearing capacity reduction coefficient of soil between piles is 0.75. Try to determine pile spacing in order to obtain characteristic bearing capacity of the composite foundation of 200 kPa.

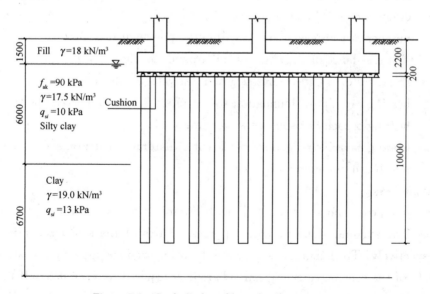

Figure 7-8 Geological profile and soil parameters

Solution (1):

According to Equations 7-35 and 7-36,

$$R_a = u_p \sum_{i=1}^{n} q_{si} l_i + \alpha q_p A_p$$
$$= \pi \times 0.55[(10 \times 5.1) + (13 \times 4.9)] + 0.5 \times 120 \times \pi \times 0.55^2/4 \text{ kN} = 212.4 \text{ kN}$$
$$R_a = \eta f_{cu} A_p = 0.3 \times 1,800 \times \pi \times 0.55^2/4 \text{ kN} = 128.22 \text{ kN}$$

The characteristic vertical bearing capacity of a single cement-soil mixed pile = minimum {212.4 128.22} =128.22 kN

Solution (2):

According to Equation 7-20,

$$f_{spk} = m \frac{R_a}{A_p} + \beta(1-m) f_{sk}$$

$$200 = m \times \frac{180}{0.237} + 0.75(1-m) \times 100 \Rightarrow m = 0.183$$

$$m = \frac{d^2}{d_e^2} = \frac{0.55^2}{(1.13 \times s)^2} = 0.183 \Rightarrow s = 1.138 \text{ m}$$

7.6 Grouting reinforcement ground

7.6.1 Introduction of grouting method

Grouting method is pouring some solidified materials, such as cement, lime or other chemical materials, into the rock and soil in a certain range under the foundation to fill the cracks and pores in the rock and soil. It can prevent the foundation from leaking, and improve the integrity, strength and rigidity of the rock and soil. In water retaining structures such as dams and dikes, grouting method is often used to build foundation seepage-proof curtain. Grouting is the main ground improvement method for hydraulic structures.

7.6.2 Classification of grouting method

According to their functions, grouting methods can be divided into the following three categories according to their functions.

1. Permeation grouting

In permeation grouting, slurry is poured into cracks of rock or pores of soil by pressure through grouting holes. The slurry replaces the gas in the pores and pore water of soil. After solidification, it binds the broken rock mass or discrete soil particles together. Therefore, permeation grouting can greatly reduce the permeability of rock and soil, and obviously improves the integrity, strength and rigidity. According to grouting materials, there are two different methods given as follows.

(1) Cement grouting and clay-cement grouting

In the cement grouting and clay-cement grouting, the cement slurry or clay-cement slurry is grouted into rock foundation to plug cracks, or into gravel foundation cover layer to fill pores. The cement grouting or clay-cement grouting is a common ground improvement method for hydraulic structures, which is used to form an impervious curtain and increase the integrity of fractured rock mass. When using this kind of particle slurry, special attention should be paid to its grouting property. In principle, as long as the particle size of the grouting material is smaller than the size of the effective pores or cracks of the grouted soil, the slurry can be grouted. However, in the grouting process, especially when the slurry concentration is high, slurry often enter the pores or cracks in two or more particles at the same time, thus blocking the grouting channel. Therefore, it is not enough to meet the conditions, but also to consider the additional influence produced by the clogging effect of particles.

(2) Chemical material grouting

A good grouting material should have good grouting property, which can control the setting time of slurry and has high strength after solidification. Meanwhile, a good grouting material should not be corroded or dissolved by water and has good durability. With the development of modern chemical industry, various chemical grouting materials with good performance have been developed. Chemical grouting materials can be divided into several categories, such as polyurethane, acrylamide, epoxy resin, methyl methacrylate, lignin, silicate and sodium hydroxide. Acrylamide materials are well groutable and the grouting process can be accurately controlled. However, they have been banned in Japan and the United States because they will pollute the air and groundwater.

2. Fracture grouting

Permeation grouting depends on the permeability of the soil, and the slurry is pumped into the pores of the soil by pressure. For cohesive soil or soft clay with small permeability, slurry can't be poured into soil pores in a short time, and permeation grouting is not applicable. In order to eliminate hidden cracks and holes in this kind of soil, fracture grouting is often used. Fracture grouting is to pump slurry into the borehole at a suitable pressure, and the soil is fractured when the pumping pressure is enough to overcome the initial stress (usually self-weight stress) of the soil layer and the tensile strength of the soil. Fractured cracks are generally perpendicular to the direction of small principal stress, which are vertical cracks. They are easy to connect with the hidden cracks and holes in the soil, so the slurry fills the hidden cracks and holes through the cracks, thus strengthening the soil.

Fracture grouting is an important method to deal with hidden dangers in dams. During the construction of cohesive soil dikes along rivers and lakes, there may be soft soil or holes in the dikes due to local pressure leakage, or internal cracks due to uneven settlement. These hidden internal damages may form hidden dangers. Fracture grouting is an effective reinforcement method when the exact crack position cannot be found.

3. Compaction grouting

Different from the above two kinds of grouting, compaction grouting is to inject thick slurry into the foundation soil through drilling. Thick slurry cannot penetrate into the pores of the soil, so the surrounding soil is compacted at the slurry outlet section to form slurry bubbles, as shown in Figure 7-9. The shape of the pulp bubble is generally cylindrical. When the diameter of the slurry bubble is small, the grouting pressure basically develops along the radial direction of the borehole, that is the horizontal direction, which makes the surrounding soil squeezed. Practice has proved that the soil can be obviously compacted within the range of 0.3-2.0 m from the slurry bubble interface. As the slurry bubble continues to expand outward, its shape may become spherical, which will generate a large upward lifting force and make the ground move. If the grouting pressure can be used reasonably to form an appropriate lifting force, the sinking building can be recovered to the required position.

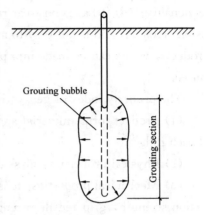

Figure 7-9 Schematic diagram of compaction grouting

Compaction grouting is suitable for strengthening soft cohesive soil, such as silt and muddy soil. However, under the condition of low permeability coefficient and poor drainage, high pore water pressure may be caused in the reinforced soft soil. In this case, in order to prevent soil damage, very low grouting rate and solidification rate must be used.

7.7 Geosynthetics reinforcement ground

Commonly used reinforcement methods include geosynthetics reinforcement method and soil nailing reinforcement method. Soil nailing is an important method of pit wall support in deep foundation pit excavation, which is described in Chapter 6. Geosynthetics is a new material widely used in geotechnical engineering, and its function is not limited to reinforcement. This section focuses on its reinforcement function, but also briefly introduces other geotechnical engineering applications.

7.7.1 Types of geosynthetics

Geosynthetic refers to the synthetic material products used in geotechnical engineering. It is made of synthetic polymers (such as plastics, chemical fibers, synthetic rubber, etc.) as raw materials, and placed in the interior, surface or between soil layers to strengthen or protect the soil. The occurrence and wide application of geosynthetics is one of the most important achievements in geotechnical engineering practice since the second half of the 20th century.

The characteristics of geosynthetics include: (a) soft texture, light weight and good overall continuity; (b) the construction is convenient, the cost is low, and the tensile strength is high; (c) their strength in all directions is almost similar without obvious directionality; (d) it has good wear resistance, corrosion resistance, durability and anti-microbial erosion, and is not easy to mildew and insect erosion; (e) the materials are factory products, which can be made into permeable materials and non-permeable materials as required.

The property index of geosynthetics mainly include the following items.

(1) Product form: material and manufacturing method, width, diameter and weight of each roll.

(2) Physical properties: mass per unit area, thickness, opening size and uniformity.

(3) Mechanical properties: tensile strength, elongation at break, tear strength, penetration strength, grip tensile strength, top breaking strength, fatigue strength, creep, and etc.

(4) Hydraulic properties: vertical and horizontal water permeability, clogging and waterproofing, and etc.

(5) Durability: UV resistance, chemical stability and biological stability.

At present, there are many kinds of geosynthetics. They can be divided into the following categories, which are shown in Figure 7-10.

(1) Geotextiles

Geotextiles are one kind of geosynthetic materials with water permeability. According to different manufacturing methods, it can be divided into woven geotextiles and non-woven geotextiles. Woven geotextile is a geotextile woven by arranging fibers or filaments in a certain direction. Nonwoven geotextile is a geotextile made by mechanical bonding, thermal bonding, or chemical bonding of thin flocs arranged randomly or directionally by short fibers or filaments. Geotextiles are generally used for drainage, filtration, reinforcement and soil isolation.

(2) Geomembrane

Geomembrane is a relatively impermeable membrane made of polymers (including asphalt). It can be divided into asphalt and polymer according to its raw materials. According to its products, it can be divided into single membrane and composite membrane. The latter is made of geotextile reinforced with fabric. Geomembrane can be widely used as seepage control material due to its relatively impermeable characteristics.

(3) Geogrid

Geogrid is a reinforced geosynthetic material formed by the combination of tensile strips in a regular grid pattern, with openings that can accommodate the embedding of filling materials. Geogrid is divided into plastic geogrid, fiberglass geogrid, polyester warp

Figure 7-10 Types of geosynthetics

knitted geogrid, and steel plastic geogrid bonded or welded by multiple composite reinforcement strips. Geogrid is mainly used for soil reinforcement.

(4) Geocell

Geocell is a honeycomb or grid shaped three-dimensional structural material composed of strips. These strips are formed by geogrids, geotextiles, or geotextile membranes with a certain thickness, which are combined and connected to each other. It is mainly used for reinforcement of road base layer or foundation, and can also be used for erosion control works.

(5) Geonet

Geonet is a two-dimensional network of geosynthetic materials formed by connecting strip components at nodes, which can be used for isolation, wrapping, drainage, and exhaust.

(6) Geomat

Geomat is a three-dimensional structure made of thermoplastic resin, also known as a three-dimensional vegetation network. The bottom layer is the foundation layer, covered with a bubble shaped expanded mesh bag, filled with fertile soil and grass seeds for plant growth.

(7) Geocomposite

Geocomposite is a geosynthetic material composed of two or more materials. It has been found that the combination of several different geosynthetics can achieve more ideal results. Currently, there are various geocomposite. For example, single-layer membrane and geotextile is combined to form composite geomembrane. Plastic drainage belt is formed by geotextile and plastic corrugated board. Geotextiles and geogrids are used as reinforcement materials in cohesive soil.

7.7.2 Functions of geosynthetics

The application of geosynthetics in geotechnical engineering mainly plays the following functions.

1. Drainage function

Some geosynthetics can form drainage channels in the soil, collect the water in the soil and discharge it under the action of water level difference. One example is the prefabricated vertical drain used in the preloading and consolidation treatment of saturated soft clay. A certain thickness of geotextile or geomat has good vertical and horizontal permeability, and can be used as a drainage facility to effectively collect and discharge water from the soil.

2. Filter function

As geotextile has good water permeability and the ability to prevent particles from passing through, it is an ideal material for use as filtermaterial. In various civil engineering projects, such as embankment dam, earth embankment, subgrade, culvert gate, retaining wall, etc., it can be used to replace the traditional gravel filter, and achieve economic benefits and good technical performance. Geotextiles used for filtration are generally non-woven geotextiles, and sometimes woven geotextiles can also be used. The basic requirements are as follows: (a) Under the action of water flow, the protected soil must not be carried away by the water flow. That is, it needs to have "soil retention" to prevent piping damage. (b) The water flow must be able to pass through the fabric plane smoothly. That is, it needs to have "water permeability" to prevent excessive seepage pressure caused by accumulated water. (c) The pore size of the fabric should not be blocked by soil particles carried by water flow. That is, it should have anti-clogging to avoid the failure of the filtration effect.

3. Separation

Separation is the placement of geosynthetic materials between two different materials or between two different soils, so that they do not mix with each other. For example, separating crushed stones from fine-grained soil, and separating soft soil from fill. Separation can produce excellent engineering and technical effects. When the structure is subjected to external loads, the separation prevents materials from mixing or losing each other, thereby maintaining its overall structure and function. The use of separation geosynthetics should be determined based on their use in engineering, and woven geotextiles have been the most commonly used. If there is a high requirement for the strength of the material, geonet or geogrid can be used as the cushion of the material. When separation and seepage prevention are required, the geocomposite of geotextile and membrane can be used. The materials used for separation must have sufficient resistance to bursting and puncture.

4. Reinforcement

Reinforcement is the embedding of geosynthetic materials (reinforcement) with high tensile strength, tensile modulus, and large surface friction into the soil. Through the stress transfer of frictional resistance between the reinforcement and the surrounding soil interface, it constrains the lateral displacement of the soil under stress, thereby improving the bearing capacity or structural stability of the soil. Geosynthetic materials used for reinforcement include woven geotextiles, geotextiles, and geogrids. They are widely used in soft soil foundation reinforcement, steep slopes of embankments, and retaining walls. The geosynthetic material used for reinforcement should have good adhesion to soil and low creep resistance. At present, geogrids are the most ideal reinforcement material.

5. Seepage prevention

The almost impermeable geomembrane can achieve ideal seepage control effect, and can be used for seepage control of canals, ponds, reservoirs, earth-rock dams, gates and foundations. In recent years, it has also been widely used in landfill sites to prevent leachate from polluting groundwater. Geomembrane has good anti-seepage effect, light weight, convenient transportation, simple construction, and low cost.

6. Protection

Protection function refers to the protective effect of geosynthetics and structures or components composed of geosynthetics on soil. Geosynthetics can have protection effect on soil or water surfaces, such as preventing river bank or coastal erosion, preventing frost damage to soil, and preventing water surface evaporation or pollution from dust in the air.

It can be used for erosion prevention of river channels and harbor slopes, as well as pollution prevention of reservoirs and channels.

Questions

7-1 What is the purpose of ground improvement?

7-2 What are the main categories of ground improvement? What is the mechanism of its reinforcement of soil?

7-3 What is the composite foundation? What are the main differences between composite foundation and piles?

7-4 What are the types of composite foundations? What are the main features of the design?

7-5 What are the requirements for cushion materials for the replacement cushion method?

7-6 How to determine the width and thickness of the cushion?

7-7 What are the main characteristics of cement-soil mixed method and high-pressure jet grouting method? What soil type are they suitable for?

7-8 What is the dynamic compaction method? Explain the mechanism of dynamic compaction method for strengthening foundation.

7-9 How to determine the effective reinforcement depth of dynamic compaction method?

7-10 What are the types of preloading methods? Briefly explain the reinforcement principles of each type of method.

7-11 What is the principle of sand well preloading method? How to calculate the degree of consolidation of preloading?

7-12 How to determine the bearing capacity of the foundation after sand pile reinforcement?

7-13 Grouting method is the main method for strengthening the foundation of hydraulic structures. What are the types of grouting methods? What conditions are they applied to?

7-14 What are the commonly used types of chemical grouting? What are the main advantages and disadvantages?

7-15 What is compaction grouting? What situation is it applicable for?

Chapter 8 Special Soil Subgrade

China has a vast land, ranging from coastal areas to inland areas, and from mountainous areas to plains, where various types of soil are widely distributed. Some soil types have special components, structures, and properties due to different geographical environments, climatic conditions, geological origins, historical processes, and secondary changes during their generation. When used as the subgrade of a building, failure to pay attention to these soils' special properties may cause accidents. This chapter mainly introduces the special engineering properties of these soils which are widely distributed in China and the engineering measures that should be taken when used as subgrade.

8.1 Collapsible loess subgrade

8.1.1 Main characteristics and distribution of loess

Loess is a kind of deposit in arid and semi-arid climate conditions in Quaternary geological history. It is widely distributed in many parts of the world, accounting for about 9.3% of the total land area. The internal material composition and external morphological characteristics of loess are different from other sediments of the same period, and their geographical distribution also has certain regularity. The general characteristics of loess include: (a) The color is yellow or brownish yellow, as Figure 8-1 shown. (b) The main

Figure 8-1 The photo of loess

particle is silt and its content often accounts for more than 60%. (c) Void ratio e is large, generally ranging from 0.8 to 1.2. And the loess usually has macroscopic pores and vertical joints. (d) It is rich in calcium carbonate salts.

The loess in the world is mainly distributed in the mid latitude arid and semi-arid regions of the Northern Hemisphere. Except for South America and New Zealand, loess is rarely distributed in other regions of the southern hemisphere. The loess in Asia is the most widely distributed, with its northernmost boundary reaching 74°N and its southernmost reaching 32°N (near Nanjing). The northernmost boundary of loess distribution in Europe is 62°N, and the southern boundary is around 40°N. In North America, loess is mainly distributed in the ancient glacier front zone near 40° north latitude. In the southern hemisphere, loess is mainly distributed in the Bams Grassland area of Argentina near 40°S latitude.

Loess and loess like soil in China are mainly distributed in arid and semi-arid areas north of the Kunlun Mountains, Qinling Mountains, Mount Taishan and Mount Lushan. The primary loess is best developed in the middle reaches of the Yellow River, mainly in Shanxi, Shaanxi, southeastern Gansu, and western Henan. In addition, there are scattered distribution in Beijing, western Hebei, eastern Qinghai, Xinjiang region, Songliao Plain, Sichuan, Three Gorges, Huaihe River basin in northern Anhui, and Nanjing. The arc distribution of loess in China is not only controlled by mountain terrain, but also related to the zonal distribution of China's climate. Loess is mainly distributed in temperate regions, with annual average rainfall of 300-700 mm and rainfall less than evaporation.

The special properties of loess are given as follows.

(1) Particle size composition

The particle size composition of loess is one of the representative characteristics that distinguishes it from other Quaternary sediments. The composition of loess is uniform, characterized by a high content of silt particles (0.05-0.005 mm), with coarse silt particles (0.05-0.01 mm) accounting for over 50% and clay particles (0.25 mm particles). In short, loess is a sediment mainly composed of silt and contains a certain proportion of fine sand, extremely fine sand and clay particles.

(2) Mineral composition

The loess includes detrital minerals and clay minerals, and the former accounts for over 70%. In detrital minerals, it mainly refers to light minerals with density less than 2.90, accounting for more than 90%. Quartz is the largest, followed by feldspar, and some carbonate minerals (calcite, dolomite, etc.). The clay minerals in loess are mainly hydromica, kaolinite and montmorillonite. The presence of these minerals endows loess with characteristics such as adsorption, expansion, and contraction, which affects the engineering properties of loess. Carbonate minerals often play a role in cementation, resulting in the presence of aggregates in the natural structure of loess. After encountering water, the aggregates cemented by soluble salts are destroyed, often enhancing the collapsibility of loess. Therefore, some people believe that the presence of carbonates in loess is one of the reasons for loess collapse. Although the detrital minerals of loess like soil are mainly quartz and feldspar, the content of unstable mineral components is relatively low, and they are generally strongly weathered and related to nearby bedrock.

(3) Chemical composition

The chemical composition of loess is determined by its mineral composition. The highest content is SiO_2, because there are also aluminum silicate minerals in loess in addition to a large amount of quartz. The second highest content is Al_2O_3, as the main mineral in loess is feldspar. CaO content is also very high, because loess contains calcite.

(4) Structural construction

Loess generally has well-developed tubular pores, whose large pore sizes range from 0.5 to 1 cm. The pores are mostly filled with carbonate, and some pores are almost entirely filled with carbonate. The porosity of loess is generally from 33% to 64%. It has been proved that the porosity of loess is the fundamental reason for its collapsibility.

The formation of vertical joints in loess is produced due to two reasons: (a) The tensile stress acts on the interior of the loess in the vertical direction under the action of gravity. (b) There are numerous vertical pores in the loess. They weaken the horizontal resultant force and make it easy to generate cracks along the vertical direction. In the slope areas of loess areas, collapse often occurs along the joint surface, which forms steep cliffs and sometimes leads to large-scale landslides.

Loess does not have bedding. Although the grain size of loess varies in thickness in the vertical profile, such changes do not cause obvious layering alternation, which is different from general bedding. Loess soil has fewer pores than loess, but its bedding is very clear.

Due to the wide geographical distribution and the long age spanned by the formation time, the properties of loess are very different. In order to better understand the speicail properties of the loess, the loess has been classified from different views. At present, there are two kinds of classification. One is to classify loess into old loess and new loess according to the geological age of formation. Old loess refers to loess formed in the early Pleistocene (referred to as Q_1 loess or Wucheng loess) and loess formed in the middle Pleistocene (Q_2 loess or Lishi loess). New loess refers to loess (Q_3 loess or Malan loess) formed in late Pleistocene and loess (Q_4 loess) formed in Holocene. There are some Q_4 loess with short sedimentation age, uneven soil quality, loose structure, high compressibility, low bearing capacity and great difference in collapsibility in loess, which is called newly accumulated loess Q_4^2 in order to attract the attention of engineering design. It is generally believed that Q_1, Q_2 and Q_3 loess is primary loess, with wind as the main factor. And the newly accumulated Q_4 loess is the secondary loess, with water as the main factor. Obviously, the longer the loess is formed, the deeper the stratum position, the higher the compactness of loess, the better the engineering properties, and the collapsibility decreases until there is no collapsibility. Another classification system is based on the collapsibility of loess after meeting water, which is divided into collapsible loess and non-collapsible loess.

8.1.2 Collapsibility of loess and its influencing factors

Collapse of loess refers to a kind of deformation with large subsidence and fast subsidence caused by structural damage after loess is soaked by water. This deformation is different from the compression deformation of ordinary fine-grained soil after soaking saturation. The compressibility of ordinary fine-grained soil only increases slightly after soaking saturation, unlike the rapid and large-scale collapse deformation of collapsible loess after soaking.

Collapse of loess is a complex geological, physical and chemical process. There have been various hypotheses and theories concerning the collapsing deformation of loess soils, e.g. under-compaction theory, salt dissolution assumption, clay-colloidal shortage hypothesis, water-film wedge action and capillary effect. However, a unified theory that can fully explain all the phenomena and essence of collapse has not yet been obtained. The following is a brief introduction to several widely recognized and reasonable hypotheses.

The under-compaction theory believes that in arid and less rainy climates, water constantly evaporates during the sedimentation process of loess. And salt precipitates between soil particles, which forms solidified cohesive forces. When the soil moisture is not high, the overlying soil layer is insufficient to overcome the solidified cohesive forces formed in the soil, resulting in an under-compaction state. Once soaked by water, the solidified cohesive forces disappear, which leads to settlement.

The salt dissolution assumption suggests that loess collapse is due to the presence of a large amount of easily soluble salt in the loess. When the water content in loess is low, soluble salts are in a microcrystalline state and attach to the surface of particles, which plays a cementing role. After being soaked in water, soluble salts dissolve and the cementation is lost resulting in collapse. However, the salt dissolution hypothesis cannot explain all collapsible phenomena, such as the relatively low content of soluble salts in collapsible loess in China.

The structural theory believes that the fundamental cause of loess collapse is its special granular overhead structure system. This structural system is formed by the interconnection of skeletal particles composed of aggregates and debris (Figure 8-2), which contains a large number of elevated pores. The connection strength between particles is formed under arid and semi-arid conditions, originating from the compaction of the overlying soil weight and capillary pressure formed at the contact between particles with a small amount of water. Under the action of water and external loads, this structural system will inevitably lead to a decrease in connection strength, failure of connection points, and loss of stability of the entire structural system.

Although there are different viewpoints on explaining the reasons of loess collapse, it can be summarized into two aspects: external and internal factors. The external reasons of collapsibility of loess are water immersion and load action, while the structural characteristics and material composition of loess are the internal reasons of collapsibility.

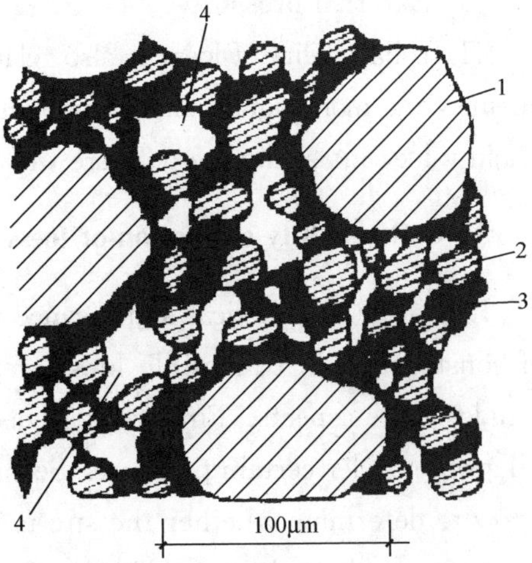

Figure 8-2 Schematic diagram of loess structure
1-sand particles; 2-silt particles; 3-cement; 4-large voids

The factors affecting the collapsibility of loess can be given as follows.

(1) Material composition of loess

The amount and composition of cement in loess, as well as the composition and distribution of particles, have a significant impact on the structural characteristics and collapsibility of loess. If the content of cement is high, it can surround the skeleton particles, resulting in a dense structure. The content of clay particles, especially those with strong bonding ability less than 0.001 mm, is high, and their uniform distribution between the frameworks also plays a role as a binder, reducing collapsibility and improving mechanical properties. On the contrary, the number of particles with a particle size greater than 0.05mm increases, and the binder is mostly distributed in a thin film shape. The skeleton particles mostly come into direct contact with each other, resulting in a loose structure, reduced strength, and enhanced collapsibility. The collapsibility of loess in China shows a decreasing trend from northwest to southeast, which is consistent with a decrease in sand content and an increase in clay content from northwest to southeast. In addition, the salts in loess and their presence state also have a direct impact on collapsibility. For example, when calcium carbonate, which is difficult to dissolve and has cementation effect, collapsibility decreases, but the higher the content of gypsum and other soluble salts such as carbonate, sulfate, and chloride, the stronger the collapsibility.

(2) Physical properties of loess

The physical properties of loess are related to the physical properties of soil such as pore ratio and water content. The larger the natural pore ratio or the smaller the natural water content, the stronger the collapsibility. Loess with a saturation $S_r \geqslant 80\%$ is called saturated loess, and its collapsibility has deteriorated. When the natural moisture content is the same, the collapsible deformation of loess increases with the increase of humidity.

(3) External pressure

The collapsibility of loess is also related to the external pressure. The greater the external pressure, the more significant the amount of collapse. However, when the pressure exceeds a certain value, increasing the pressure will actually reduce the amount of collapse.

8.1.3 Collapsibility evaluation of loess

It is of great practical significance to correctly evaluate the collapsibility of subgrade for construction in collapsible loess area. The collapsibility evaluation of loess generally includes three aspects. Firstly, it is necessary to find out whether the loess soil layer is collapsible under certain pressure. Secondly, if it is a collapsible loess soil layer, it is necessary to determine whether the site is self-weight collapsible or non-self-weight collapsible, because the collapse accident of self-weight collapsible loess subgrade after being soaked by water is more serious than that of non-self-weight collapsible loess subgrade. Finally, the collapsibility grade of collapsible loess subgrade is judged, that is, the severity of collapsibility is judged according to the collapsibility type of the site and the amount of collapsibility deformation that may occur when the subgrade is fully immersed in water under the specified pressure.

The evaluation criteria for collapsibility of loess subgrade vary from country to country. The evaluation method introduced below is based on *Code for Building Construction in Collapsible Loess Regions* (GB 50025—2018) in China.

1. Collapsibility coefficient and evaluation of loess collapsibility

Whether loess has collapsibility and the collapsibility level should be judged according to the collapsibility coefficient δ_s measured by indoor collapsibility test.

The collapsibility coefficient δ_s is obtained through the confined compression test the same as general undisturbed soil. Place undisturbed soil samples into a confined compression instrument and load one by one step pressure. After reaching the specified pressure p and settling stably, measure the height of soil sample. Then the soil sample is saturated with water. After the additional subsidence is stable, the height of the soil sample after water immersion is measured (Figure 8-3). The collapsibility coefficient δ_s can be calculated according to the following equations.

$$\delta_s = \frac{h_p - h'_p}{h_0} \quad (8\text{-}1)$$

or

$$\delta_s = \frac{e_p - e'_p}{1 + e_0} \quad (8\text{-}2)$$

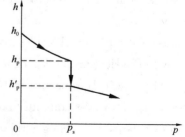

Figure 8-3 Compression curve of loess under pressure

where

h_0 ——initial height of soil sample, mm;

h_p ——height of soil sample after compression and

stability under pressure p, mm;

h'_p ——height after additional subsidence stability under the action of soil sample immersion (saturation), mm;

e_0 ——initial pore ratio of soil sample;

e_p ——pore ratio of soil sample after settlement and stability under the action of pressure p;

e'_p ——pore ratio after immersion (saturation) subsidence stability.

It is not difficult to see from Equation 8-1 that the collapsibility coefficient δ_s is the additional strain of the soil sample due to water saturation. Obviously, if the tested δ_s are small, the collapsibility is weak; The δ_s depends not only on the collapsibility of the soil, but also on the pressure p when it is immersed in water. In the test, the pressure p used to determine the collapsibility coefficient is the actual pressure of loess in the subgrade. The *Code for Building Construction in Collapsible Loess Regions* (GB 50025—2018) stipulates that: from the bottom of the subgrade (if the elevation of the basement is uncertain, it will be1.5 m below the ground), the p value of the soil layer within 10 m of the basement shall be 200 kPa. From below 10 m to the top surface of non-collapsible soil layer, the saturated dead weight pressure of the overlying soil shall be applied (when it is greater than 300 kPa, 300 kPa shall still be applied). However, if the basement pressure is greater than 300 kPa, the δ_s of loess should be judged according to the measured value under the actual pressure. For the newly deposited loess with high compressibility, the soil layer within 5 m of the basement should use 100-150 kPa pressure, and the top surface of the non-collapsible loess layer below 5-10 m and 10 m should use 200 kPa and the saturated dead weight pressure of the overlying soil layer respectively.

The main purpose of collapsibility coefficient δ_s in engineering is to judge the collapsibility of loess. When the δ_s is lower than 0.015, it is designated as non-collapsible loess. When it is not lower than 0.015, it is designated as collapsible loess. The *Code for Building Construction in Collapsible Loess Regions* (GB 50025—2018) divided collapsibility degree of the loess into the following three types according to the δ_s:

(1) when $0.015 \leqslant \delta_s \leqslant 0.03$, the collapsibility is slight;

(2) when $0.03 < \delta_s \leqslant 0.07$, the collapsibility is moderate;

(3) when $\delta_s > 0.07$, the collapsibility is strong.

The above method can only measure the collapsibility coefficient under a certain specified pressure, and sometimes it is necessary to determine the initial collapsibility pressure p_{sh} in engineering. At this time, it is necessary to find out the relationship between different immersion pressures p and collapsibility coefficient δ_s. Therefore, it can be determined by single-line or double-line collapsibility test of indoor compression test.

Single-line collapsibility test means that at least five ring cutter samples are taken at the same depth at the same soil sampling point. All these five samples are loaded step by step under natural water content, and are respectively added to different specified pressures. After the settlement is stable, they are saturated with water until the additional settlement is stable. According to Equation 8-1, the collapsibility coefficient δ_s corresponding to each pressure p can be calculated.

Double-line collapsibility test refers to taking two ring cutter samples at the same depth at the same soil sampling point. One sample is loaded step by step under natural water content. And the other is loaded with the first load under natural water content, and then it is immersed in water after sinking and stability. And then load step by step, as shown in Figure 8-4(a). Taking the difference between the subsidence of the two curves under the same pressure as the collapsibility, the corresponding collapsibility coefficient values under various pressures can also be calculated according to Equation 8-1, and the relationship curve as shown in Figure 8-4(b) can be drawn. The pressure corresponding to $\delta_s = 0.015$ on this curve p-δ is taken as the initial pressure of collapsibility p_{sh}. The p_{sh} is a very useful index. When the stress in the subgrade (the sum of self-weight stress and additional stress) is less than p_{sh}, the amount of collapse caused by immersion is very small, which can be considered as general non-collapsible subgrade. Generally speaking, the collapsibility measured by double-line method is less than that by single-line method.

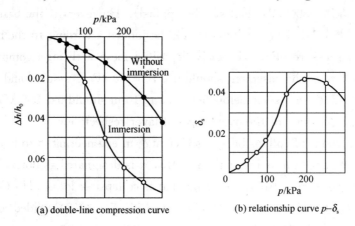

(a) double-line compression curve   (b) relationship curve p-δ_s

Figure 8-4 Immersion compression test curve

2. Collapse type evaluation of construction sites

Engineering practice shows that the collapse of self-weight collapsible loess site causes more accidents than that of non-self-weight collapsible loess site. The former site is more harmful to buildings. Therefore, it is very important to investigate the construction site before design and correctly classify the collapse types of the site. There are two ways to classify the collapse types of building sites: one is to judge according to the measured value of self-weight collapse Δ'_{zs} in the field trial pit immersion test; The other is based on the calculated value of self-weight collapsibility Δ_{zs} accumulated by indoor loess collapsibility test. Although the first method is more accurate and reliable, it is time-consuming and water-consuming. And sometimes it is not easy to do because of various conditions. Therefore, the test pit immersion test should be adopted for the important buildings in the newly-built area, yet the general buildings can be calculated according to their own weight collapse.

(1) Calculation of self-weight collapse

In order to calculate the self-weight collapse, the self-weight collapsibility coefficient δ_{zs} should be determined first. The determination method of δ_{zs} is the same as δ_s. The calculation equation can be given as follows:

$$\delta_{zs} = \frac{h_z - h'_z}{h_0} \tag{8-3}$$

where

h_z ——height of subsidence stability when pressurized to saturated dead weight pressure of overlying soil, mm;

h'_z ——height of additional subsidence stability after soaking, mm.

According to the measured self-weight collapsibility coefficient δ_{zs} of soil layers at different depths, the calculated self-weight collapse Δ_{zs} of this site can be calculated by the following equation:

$$\Delta_{zs} = \beta_0 \sum_{i=0}^{n} \delta_{zsi} h_i \tag{8-4}$$

where

δ_{zsi} ——the self-weight collapsibility coefficient of the ith-layer soil;

h_i ——the thickness of the ith layer of soil, mm;

β_0 ——a parameter related to the different regions (Table 8-1).

The parameter of β_0 Table 8-1

Regions	β_0
Region ① (Longxi region)	1.5
Region ② (Longdong-Northern Shaanxi- Western Shanxi)	1.2
Region ③ (Guanzhong region)	0.9
Other regions	0.5

(2) Collapse type classification of site

The collapse types of the construction sites can be classified according to the tested self-weight collapse Δ_{zs} of the indoor collapsibility test or the measured self-weight collapse Δ'_{zs} by the field immersion test:

when Δ_{zs} or $\Delta'_{zs} \leqslant 70$ mm, it should be designated as a non-self-weight collapsible loess site;

when Δ_{zs} or $\Delta'_{zs} > 70$ mm, it should be designated as self-weight collapsible loess site.

When there is a contradiction between the measured value of self-weight collapsibility Δ_{zs} and the calculated value Δ'_{zs}, it should be judged according to the measured value of

Chapter 8 Special Soil Subgrade

self-weight collapsibility Δ_{zs}.

3. Collapsibility grade classification of loess subgrade

when evaluating the collapsibility grade of the subgrade of a building, we should consider not only the collapsibility caused by self-weight, but also the collapsibility caused by additional stress in subgrade. The collapsibility grade of loess subgrade is divided according to the calculated subgrade collapsibility Δ_s and the calculated site self-weight collapsibility Δ_{zs}.

(1) The calculated value of subgrade collapse Δ_s

When the collapsible loess subgrade is saturated by water and sinks stably, the calculated value of the subgrade collapse Δ_s shall be calculated according to the following equation:

$$\Delta_s = \sum_{i=1}^{n} \beta \delta_{si} h_i \tag{8-5}$$

where

δ_{si} —— the collapsibility coefficient of the ith layer of soil;

h_i —— the thickness of the ith layer of soil, mm;

β —— parameter considering the possibility of subgrade soil wetting by water and lateral extrusion, which is given in Table 8-2.

The parameter of β   Table 8-2

Position and depth		β
0-5 m under the base of the foundation		1.5
5-10 m under the base of the foundation	non-self-weight collapsible loess site	1.0
	self-weight collapsible loess site	β_0 in the site and not less than 1.0
10 m below base of the foundation to the top of non-collapse soil	non-self-weight collapsible loess site	1.0 if the site is Region ① and Region ②; β_0 for other regions
	self-weight collapsible loess site	β_0 in the site region

(2) Classification of collapsibility grade

According to the above calculated subgrade collapse Δ_s and site self-weight collapse Δ_{zs}, the collapse grade of collapsible loess subgrade is divided into four grades: I (slight), II (medium), III (serious) and IV (very serious), as shown in Table 8-3.

Classification of loess subgrade collapsibility grade (Unit: mm)   Table 8-3

Δ_s	Collapsibility grade		
	Non-self-weight collapsible loess site	Self-weight collapsible loess site	
	$\Delta_{zs} \leqslant 70$	$70 < \Delta_{zs} \leqslant 350$	$\Delta_{zs} > 350$
$\Delta_s \leqslant 300$	I (slight)	II (medium)	
$300 < \Delta_s \leqslant 700$	II (medium)	II (medium) or III (serious) *	III (serious)
$\Delta_s > 700$	II (medium)	III (serious)	IV (very serious)

* When the calculated subgrade collapse $\Delta_s > 600$ mm and the calculated self-weight collapse $\Delta_{zs} > 300$ mm, it can be classified as Grade III. And other situations can be classified as Grade II.

8.1 Collapsible loess subgrade

Example 8-1:

For a loess construction site in Hebei, a soil sample is taken every 1 m during engineering investigation. The δ_{zs} and δ_s of each soil sample is measured as shown in Table 8-4. Try to determine the collapsibility type of the site and the collapsibility grade of the subgrade.

δ_{zs} and δ_s for soil samples   Table 8-4

Soil depth /m	0.5	1.5	2.5	3.5	4.5	5.5	6.5	7.5	8.5	9.5	10.5
δ_{zs}	0.01	0.014	0.02	0.017	0.05	0.01	0.02	0.015	0.04	0.002	0.002
δ_s	0.045	0.038	0.052	0.027	0.056	0.048	0.040	0.036	0.025	0.012	0.011

Note: The soil with δ_{zs} or $\delta_s < 0.015$ belongs to non-collapsible soil layer.

Solution:

(1) Determine site collapse type

Calculate the self-weight collapse Δ_{zs} from the natural ground to the top surface of the non-collapsible loess layer below.

$$\Delta_{zs} = \beta_0 \sum_{i=1}^{n} \delta_{zsi} h_i$$

Then $\Delta_{zs} = 0.5 \times (0.02 + 0.017 + 0.05 + 0.02 + 0.015 + 0.04) \times 1000 = 81.0$ mm > 70 mm.

Therefore, the site should be classified as a self-weight collapsible loess site.

(2) Subgrade collapsibility grade determination

Calculate the loess subgrade collapse Δ_s according to Equation 8-5:

$$\Delta_s = \sum_{i=1}^{n} \beta \delta_{si} h_i$$

$\Delta_s = 1.5 \times (0.5 \times 0.038 + 0.052 + 0.027 + 0.056 + 0.048 + 0.5 \times 0.040) \times 1000 + 1.0 \times (0.5 \times 0.040 + 0.036 + 0.025) \times 1,000 = 414$ mm

Based on Table 8-3, the collapsible loess subgrade can be classified as Grade II (medium).

8.1.4 Engineering measures for collapsible loess subgrade

When building in collapsible loess area, the subgrade should meet the requirements of bearing capacity, collapsible deformation, compressive deformation and stability. According to the characteristics of collapsible loess and the types of buildings, various types of loess subgrade treatment methods should be taken so as to prevent or control subgrade collapse and ensure the safety and normal use of buildings. The comprehensive treatments of loess subgrade mainly include subgrade treatment, waterproof methods and structural methods. Among them subgrade treatment is the main method to prevent loess collapsibility.

1. Subgrade treatment

According to the importance of buildings, the collapse possibility of subgrade being soaked by water, and the requirement level on uneven settlement during use, different requirements are specified. Based on the Code for Building in Collapsible Loess Area, buildings are divided into four categories: Grade A, B, C and D. For buildings belong to Grade A, it is required to eliminate all the collapsible amount of subgrade, or to use pile foundation to penetrate all the collapsible soil layers, or to set the subgrade on the non-collapsible loess layer. For buildings B and C, it is required to eliminate the partial collapse of the subgrade. Grade D is a secondary building, and the subgrade can be left untreated. The common subgrade treatment methods for collapsible loess subgrade are given in Table 8-5. These methods are widely used in engineering practice and obtain rich experience in design and construction.

Common treatment methods for collapsible loess subgrade Table 8-5

Method	Application conditions	Thickness of collapsible loess that can be treated /m
Cushion method	Above the groundwater level, local or whole treatment	1-3
Dynamic compaction	Above the groundwater level, collapsible loess with $S_r \leqslant 60\%$, local or whole treatment	3-12
Compaction method	Above the groundwater level, collapsible loess with $S_r \leqslant 65\%$	5-15
Pre-immersion method	In the self-weight collapsible loess site, the subgrade collapsibility grade is Ⅲ or Ⅳ, eliminate all collapsibility of loess layer 6 m below the ground	More than 6 m, it should be treated by cushion or other methods
Other methods	It is proved to be effective by research or engineering practice	

2. Waterproof methods

The purpose of waterproof methods is to eliminate the external reasons of collapsible deformation of loess, so as to ensure the safety and normal use of buildings. One of the important measures used is to do good waterproofing and drainage of buildings during construction. Some basic waterproof methods include: do a good job in site leveling and drainage system to prevent water from accumulating on the ground; compacting the topsoil around the building; and properly dispersing water to prevent rainwater from directly infiltrating into the subgrade. The main water supply and drainage pipes should have a certain protective distance from the house. Leak detection facilities shall be provided to prevent water gushing from soaking local subgrade soil and etc.

3. Structural methods

Structural measures are complementary to the first two measures. For some buildings whose subgrades are not treated or only partially collapsed after treatment, in addition to waterproof methods, structural measures should be taken to reduce the uneven settlement of buildings or make the structure adapt to the collapse deformation of the subgrade.

4. Construction measures and use maintenance

The construction measures and use maintenance of collapsible loess subgrade should be reasonably arranged according to the characteristics and design requirements of the soil, in order to prevent construction water and site rainwater from flowing into the building subgrade and causing collapse. During use, buildings and pipelines should be regularly maintained and repaired to ensure the effective implementation of waterproof measures and prevent the foundation from soaking and sinking.

8.2 Expansive soil subgrade

8.2.1 Characteristics of expansive soil and its harm to buildings

Expansive soil is also a very important regional special soil, with its clay particles mainly composed of hydrophilic minerals. Meanwhile, it has marked deformation characteristics of water absorption swelling and water loss contraction. As we all know, the general cohesive soil also has the characteristics of swelling and contraction, but its amount is not large and it has no great influence on the project. However, the periodic deformation of swelling-contraction-reswelling of expansive soil are very obvious, which often harms to engineering. This makes expansive soil distinguished from general cohesive soil in engineering and treated as special soil.

1. General characteristics and distribution of expansive soil

In the natural state, the liquid index of expansive soil is often less than zero, and it is in a hard or hard-plastic state. The void ratio e of expansive soil is generally 0.6-1.1. The expansive soil usually has low compressibility with reddish brown, yellow, white or other colors. In the past, the characteristics of this kind of soil were not well understood, and engineers often mistakenly thought that it could be used as a good natural subgrade for its high strength and low compressibility. However, practice has proved that this kind of soil is potentially destructive to engineering construction, and it is very difficult to control once an engineering accident occurs.

Crack development is an important characteristic of expansive soil. There are three common cracks: vertical, oblique and horizontal. Vertical cracks are often exposed to the ground. The width of cracks gradually pinches out with the increase of depth, and the cracks are often filled with gray-green or gray-white clay.

Expansive soil is widely distributed in China. According to the available data, there are expansive soils in more than 20 provinces, autonomous regions and municipalities, such as Guangxi, Yunnan, Hubei, Anhui, Sichuan, Henan and Shandong. The same is true in foreign countries, such as the United States, where expansive soil is found in 40 of the 50 states. In addition, expansive soil is distributed in India, Australia, South America, Africa and the Middle East. At present, the engineering problem of expansive soil has become a worldwide research topic.

2. Main factors affecting the swelling and shrinkage characteristics of expansive soil

The mechanism of expansive soil with swelling and contraction characteristics is very complicated, which belongs to the theoretical and practical problems of unsaturated soil. The obvious swelling and contraction characteristics of expansive soil can be attributed to the internal mechanism and external factors of expansive soil.

The internal mechanism affecting the swelling and shrinking properties of expansive soil mainly refers to mineral composition and microstructure. Experiments show that expansive soil contains a large number of active clay minerals, such as montmorillonite and illite, especially montmorillonite, which has a large specific surface area and has great suction to water at low water content. The content of montmorillonite in soil directly determines the strength of soil swelling and contraction properties. In addition to mineral components, the spatial connection of these mineral components also affects their swelling and contraction properties. Through scanning electron microscope analysis of expansive soil in a large number of different locations, it is known that the surface-to-surface connected aggregate is a common structural form of expansive soil, which has greater ability of water absorption swelling and water loss contraction than the granular structure.

The biggest external factor affecting the swelling and shrinking properties of expansive soil is the effect of water on expansive soil. The migration of water is the key external factor to control the swelling and shrinking properties of soil. Because only the soil has a gradient that may cause water migration, it is possible to cause soil expansion or contraction. Although a certain kind of clay has a high expansion potential, if its water content remains unchanged, there will be no volume change. It has been proved that a slight change in water content, even if only 1%-2%, is enough to cause harmful expansion.

3. The harm of expansive soil to buildings

Expansive soil which has obvious characteristics of water absorption swelling and water loss contraction has brought great harm to engineering construction. It causes a large number of light houses to crack and tilt, destroys highway subgrade, and causes landslides on banks and cutting. In 1973, the American Civil Engineering Society reported that the loss caused by expansive soil in the United States was at least $2.3 billion. In China, according to incomplete statistics, all kinds of industrial and civil shallow light structures built in expansive soil areas are damaged or destroyed due to the swelling and contraction of subgrade soil, resulting in economic losses of tens of billions of yuan every year. The railway line passing through the expansive soil area accounts for about 15%-25% of the total length of the railway, and various diseases caused by expansive soil are very serious. The direct renovation cost is more than several hundred million yuan every year. Due to the above situation, the engineering problem of expansive soil has attracted great attention from academic research and engineering practice. The research results show that the destruction of houses built on expansive soil subgrade has the following characteristics.

(1) The cracking and damage of buildings are generally characterized by regional groups. The damage of low-rise, light-weight and masonry structures is the most serious, because these buildings are light in weight and have low structural stiffness. Meanwhile, the foundation has a shallow depth, and the subgrade soil is easily affected by external environmental changes, resulting in swelling and shrinkage deformation.

(2) The building is subjected to bending and torsion in both vertical and horizontal directions, so it cracks at the corner of the building first. There are straight and inverted splayed cracks and X-shaped cross cracks on the wall (Figures 8-5a and c). The external longitudinal wall subgrade is subjected to the vertical shearing force and lateral horizontal thrust generated during the expansion of the subgrade, which causes the subgrade to move out and produce horizontal cracks, accompanied by horizontal displacement (Figure 8-5b).

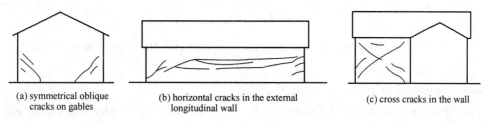

(a) symmetrical oblique cracks on gables (b) horizontal cracks in the external longitudinal wall (c) cross cracks in the wall

Figure 8-5　Wall cracks

(3) The subgrade deformation of buildings on the sloping land is not only vertical, but also horizontal, so the damage is more common and serious than that on the flat land.

8.2.2　Characteristic parameters of expansive soil

In order to distinguish expansive soil and evaluate its swelling and shrinkage, the following parameters are commonly used.

(1) Free swelling ratio δ_{ef}

The prepared ground and dried soil samples are injected into a soil measuring cup through a neckless funnel, and their volumes are measured. Then poured into a measuring cylinder filled with water, and their volumes are measured after fully absorbing water, expanding and stabilizing. The percentage of the ratio of the increased volume to the original volume is called the free swelling ratio.

$$\delta_{ef} = \frac{V_w - V_0}{V_0} \times 100\% \qquad (8\text{-}6)$$

where

V_0 ——the original volume of dry soil sample, that is, the volume of soil measuring cup, ml;

V_w ——the volume of soil sample after swelling and stability in water, which is measured by graduated cylinder, ml.

The free swelling ratio δ_{ef} indicates the expansion characteristic of dry soil particles without structural force and external pressure. It can reflect the mineral composition and content of soil. Generally, this parameter can only be used to evaluate the swelling potential of expansive soil. It cannot reflect the swelling and shrinking deformation of undisturbed soil, nor can it be used to quantitatively evaluate the swelling and shrinking range of subgrade soil.

(2) Swelling ratio δ_{ef} and swelling pressure p_e

The swelling ratio δ_{ef} indicates the percentage of the increased height of undisturbed soil or disturbed soil sample to the original height in the confined compression instrument under a certain pressure after being soaked and expanded stably, which is expressed as

$$\delta_{ep} = \frac{h_w - h_0}{h_0} \times 100\% \qquad (8\text{-}7)$$

where

h_w ——the height of the soil sample after swelling and stability in water under a certain load, mm;

h_0 ——original height of soil sample, mm.

The swelling ratio under different pressures can be used to calculate the actual expansion or shrinkage deformation of the subgrade. The swelling ratio under 50 kPa pressure can be used to calculate the subgrade deformation and classify the swelling and contraction grades of the subgrade.

Figure 8-6 shows the relationship between the swelling ratio δ_{ef} and the pressure p. The intersection point of this curve with the abscissa is defined as the swelling pressure p_e. The swelling pressure indicates the maximum internal stress of undisturbed soil sample or disturbed soil sample due to immersion expansion when the volume is constant.

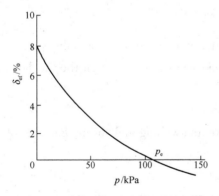

Figure 8-6 Swelling ratio-pressure curve

Swelling pressure is a very useful index when selecting foundation type and subgrade pressure. In design, if we want to eliminate swelling deformation, we should make the base pressure close to the swelling pressure.

(3) Linear shrinkage ratio δ_{sr} and shrinkage coefficient λ_s

The shrinkage of expansive soil can be expressed by linear shrinkage ratio δ_{sr} and shrinkage coefficient λ_s.

Linear shrinkage ratio δ_{sr} refers to the percentage ratio of vertical shrinkage deformation to the original height of the ring cutter soil sample dried or air-dried under natural humidity, which can be expressed as

$$\delta_{sri} = \frac{h_0 - h_i}{h_0} \times 100\% \tag{8-8}$$

where

h_0 ——original height of soil sample, mm;

h_i ——height of soil sample at a certain water content ω_i, mm.

Figure 8-7 shows the shrinkage curve based on the linear shrinkage ratio and the corresponding water content at different times. It can be seen that with the evaporation of water content, the height of soil sample gradually decreases and δ_{sr} increases. The ab part of the figure is a straight-line contraction section, and the bc part is a curve contraction transition section. After the point c, although the water content continues to decrease, the volume contraction has basically stopped.

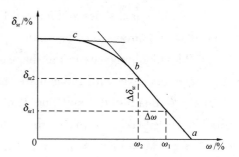

Figure 8-7　Shrinkage curve

The shrinkage coefficient λ_s can be obtained by using the linear shrinkage section, which is defined as the change value of the vertical linear shrinkage ratio corresponding to each 1% reduction of water content in the linear shrinkage stage. It can be expressed as:

$$\lambda_s = \frac{\Delta \delta_{sr}}{\Delta \omega} \tag{8-9}$$

where

$\Delta \omega$ ——the difference of water content between two points in the linear shrinkage stage, %;

$\Delta \delta_{sr}$ ——the difference of vertical linear shrinkage ratio corresponding to the difference of water content at two points in the linear shrinkage stage, %.

Shrinkage coefficient and swelling ratio are two main parameters in deformation calculation of expansive soil subgrade.

According to its landscape and topographic conditions, the site can be divided into two categories, namely flat site and sloping site. Flat site refers to the top zone with a terrain slope less than 5° or a terrain slope of 5°-14°, but the horizontal distance from the slope shoulder is more than 10 m. Slope site refers to the terrain slope greater than 5° or a terrain slope less than 5°, but with a local terrain elevation difference greater than 1 m within the same building. Attention should be paid to differential treatment in subgrade design.

The main index parameters for degerming the expansive soil in site is engineering geological characteristics and free swelling ratio. According to the *Technical Code for Building in Expansive Soil Region (GB 50112—2013)*, any site with the following engineering geological characteristics and failure modes, and cohesive soil with free swelling ratio greater than 40% should be evaluated as expansive soil.

(1) Cracks are developed and usually accompanied with glossy surface and scratches. Some cracks are filled with variegated clay such as gray and gray-green, which is hard or hard-plastic under natural conditions.

(2) Most of them are exposed in terraces, piedmont and hilly areas on the edge of the basin with gentle terrain and no obvious natural scarps.

(3) Shallow landslides and ground fissures are common, and the newly excavated pit (groove) wall is prone to collapse.

(4) Most buildings are in the shape of "invertedoctagonal", "X" or horizontal cracks, which open and close with the climate change.

When evaluating the swelling and shrinkage grade of expansive soil subgrade, it should be classified according to the influence degree of swelling and shrinkage deformation of subgrade on low-rise masonry buildings. This is because the basement pressure of light structure is small, and the swelling and shrinkage deformation is large, which is easy to cause structural damage. Therefore, the *Technical Code for Building in Expansive Soil Region (GB 50112—2013)* stipulates that the subgrade deformation calculated using the swelling ratio of soil measured under 50 kPa pressure (corresponding to the basement pressure of a masonry structure) is used as the standard for classifying swelling and shrinkage grades, as Table 8-6 shows.

Swelling and shrinkage grade classification of expansive soil subgrade   Table 8-6

Subgrade deformation s_c /mm	Grade
$15 \leqslant s_c < 35$	I
$35 < s_c \leqslant 70$	II
$s_c \geqslant 70$	III

8.2.3 Key points for the foundation design of expansive soil site

1. Subgrade design requirements

According to the importance, scale, functional requirements and engineering geological characteristics of the buildings on the expansive soil site, and the degree to which the water change in the soil may cause damage to the buildings or affect the normal use, the subgrade and foundation design are divided into three grades: A, B and C, as shown in Table 8-7.

8.2 Expansive soil subgrade

Design grade of expansive soil site Table 8-7

Design grade	Types of buildings and subgrades
A	(1) Large and important industrial and civil buildings. (2) Wet workshops with large water consumption during use, chimneys, furnaces, kilns and cold storage with negative temperature that bear high temperature for a long time. (3) Buildings with high temperature, high pressure, flammability and explosion that are strict with the subgrade deformation or sensitive to the reciprocating deformation of the subgrade. (4) Important buildings located on slopes. (5) Low-rise buildings on expansive soil subgrade with swelling and contraction Grade Ⅲ. (6) retaining structure with a height greater than 3 m and deep subgrade pit engineering with a depth greater than 5 m
B	Industrial and civil buildings other than Grade A and Grade C
C	(1) Secondary buildings. (2) Buildings on the expansive soil subgrade with flat site, simple subgrade conditions and uniform load, and the swelling and contraction grade is Grade I

According to the subgrade and foundation design grade, together with the influence degree of subgrade swelling and shrinkage deformation on the superstructure under long-term action, the subgrade design shall meet the following requirements.

(1) The subgrade should meet the bearing capacity requirement.

(2) Buildings with subgrade and foundation design grade A and B shall be designed according to the subgrade deformation requirement.

(3) The buildings (such as high-rise buildings, towering structures, retaining structures) built on or near slopes and other projects that are often subjected to horizontal loads should check their stability. The effect of horizontal swelling pressure should be considered during checking.

2. Buried depth of subgrade

Considering that the topsoil has been affected by the cyclic swelling, shrinking and drying for a long time, cracks are developed in the soil. The strength parameters of the soil, especially the cohesion, is significantly reduced. The shallow landslide on the slope often occurs within 1.0 m below the surface, which is a highly active zone. Therefore, the buried depth of the subgrade of the building should not be less than 1.0 m.

When the building has special requirements for deformation, it should be determined by calculating the swelling and shrinkage deformation of the subgrade. For multi-storey buildings on flat ground, when the buried depth of subgrade is the main prevention measure, the buried depth of subgrade should not be less than the depth of the layer with sharp atmospheric influence. For a slope with a toe of 5°-14°, when the horizontal distance from the outer edge of the subgrade to the shoulder of the slope is 5-10 m, the buried depth of the subgrade (Figure 8-8) can be determined according to Equation 8-10.

$$d = 0.45 d_a + (10 - l_p)\tan\beta + 0.3 \tag{8-10}$$

where

d ——buried depth of foundation, m;

d_a ——atmospheric influence depth, m;

β ——slope angle, °;

l_p ——horizontal distance from outer edge of subgrade to slope shoulder, m.

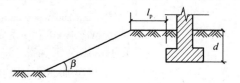

Figure 8-8 Foundation buried depth on slope

3. Evaluation of subgrade bearing capacity

Equation 3-7 and Equation 3-12 in Chapter 3 can be used to evaluate the subgrade bearing capacity of expansive soil. The modified characteristic value of subgrade bearing capacity shall be calculated according to Equation 8-11.

$$f_a = f_{ak} + \gamma_m (d - 1.0) \tag{8-11}$$

where

γ_m ——the weighted average unit weight of soil above the foundation bottom surface, and the buoyant unit weight below the groundwater level should be used, kN/m³;

f_{ak} ——the characteristic value of subgrade bearing capacity, which should be determined by in-site immersion load test for important buildings, and can be determined by local experience for areas with a lot of test data and engineering experience, kPa.

4. Subgrade deformation evaluation

The deformation of expansive soil subgrade refers to swelling or shrinkage deformation. Its deformation modes are related to local climate, topography, ground moisture, groundwater movement, ground cover, tree vegetation, building weight and other factors. Under different conditions, three different deformation modes can be produced: upward deformation, downward deformation and lifting deformation. Therefore, the deformation of expansive soil subgrade should be calculated according to the actual situation. It should be calculated according to the following three situations.

(1) Swelling deformation

The swelling deformation of the expansive soil subgrade should be calculated when the natural water content of the subgrade soil at a distance of 1 m from the ground is equal to or close to the minimum water content, or when the ground is covered and there is no possibility of evaporation, or the subgrade is often soaked by water during the use of buildings.

(2) Shrinkage deformation

The shrinkage deformation of the expansive soil subgrade should be calculated when the natural water content of subgrade soil at 1 m away from the surface is more than 1.2 times the plastic limit water content, or the subgrade directly affected by high temperature.

(3) Swelling and shrinkage deformation

In other cases, the deformation of the expansive soil subgrade should be calculated according to the swelling and shrinkage deformation.

The method to calculate the subgrade deformation can still use the Layered summation method. The above three different deformation modes are introduced as follows.

(1) Swelling deformation of subgrade soil s_e (mm)

$$s_e = \varphi_e \sum_{i=1}^{n} \delta_{epi} h_i \qquad (8\text{-}12)$$

where

φ_e —— the empirical coefficient for calculating swelling deformation, which should be determined according to local experience. If there is no local experience, 0.6 can be used for buildings with three floors and below;

δ_{epi} —— the swelling ratio of the ith layer of soil under the action of the sum of the average self-weight stress and the average additional stress of ith layer of soil;

h_i —— alculated thickness of the ith layer of soil, mm;

n —— the number of soil layers divided from the bottom of subgrade to the calculated depth z_n (Figure 8-9a), the calculated depth z_n should be determined according to the atmospheric influence depth, and if possible, it can be determined according to the immersion influence depth.

(2) Shrinkage deformation of subgrade soil s_s (mm)

$$s_s = \varphi_s \sum_{i=1}^{n} \lambda_{si} \Delta \omega_i h_i \qquad (8\text{-}13)$$

where

φ_s —— the empirical coefficient for calculating shrinkage deformation, which should be determined according to local experience; if there is no local experience, 0.8 can be used for buildings with three floors and below;

λ_{si} —— the shrinkage coefficient of the ith layer of soil under the foundation bottom surface, which should be determined by indoor tests;

$\Delta \omega_i$ —— the average value (expressed in decimals) of possible water content changes in the ith layer of soil during the shrinkage of subgrade soil (Figure 8-9b);

n —— the number of soil layers divided from the foundation bottom to the calculated depth z_n. Within the calculated depth z_n, the average variation of water content $\Delta \omega_i$ of each soil layer (Figure 8-9b) should be calculated according to Equation 8-14 and Equation 8-15. When there is impermeable bedrock in 4 m depth below the surface, it can be assumed that the variation of water content is constant (Figure 8-9c).

$$\Delta \omega_i = \Delta \omega_1 - (\Delta \omega_1 - 0.01) \frac{z_i - 1}{z_n - 1} \quad (8\text{-}14)$$

$$\Delta \omega_1 = \omega_1 - \varphi_w \omega_p \quad (8\text{-}15)$$

where

ω_1, ω_p ——natural water content and plastic limit water content of soil at 1 m below the surface (expressed in decimal);

φ_w ——moisture coefficient of soil;

z_i ——depth of the ith layer of soil, m;

z_n ——the calculation depth of shrinkage deformation, which should be determined according to the atmospheric influence depth. When there is heat source influence, it can be determined according to the depth of heat source influence. When there is a stable groundwater level within the calculation depth, it can be calculated to 3 m above the water level.

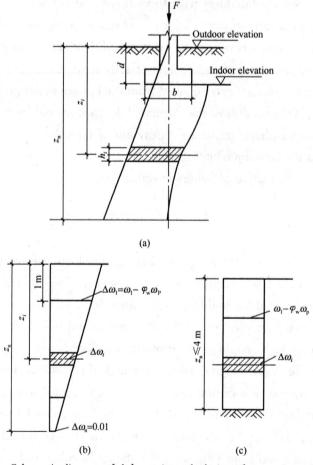

Figure 8-9 Schematic diagram of deformation calculation of expansive soil subgrade

The moisture coefficient of expansive soil φ_w refers to the ratio of the minimum water content of the soil layer at the depth of 1 m below the ground to its plastic limit under the influence of natural climate, which should be determined according to local records. If there is no such data, it can be calculated according to the equation given in the *Technical Code for Building in Expansive Soil Region* (GB 50112—2013).

The atmospheric influence depth d_a of expansive soil should be determined by the deep deformation observation, water content observation and ground temperature observation data of each climate zone. It can be used according to Table 8-8 without these data.

Atmospheric influence depth   Table 8-8

Moisture coefficient of soil φ_w	Atmospheric influence depth d_a/m
0.6	5.0
0.7	4.0
0.8	3.5
0.9	3.0

(3) Swelling and shrinkage deformation of subgrade soil s_{es} (mm)

$$s_{es} = \varphi_{es} \sum_{i=1}^{n} (\delta_{epi} + \lambda_{si}\Delta\omega_i) h_i \qquad (8\text{-}16)$$

where

φ_{es} ——the empirical coefficient for calculating the swelling and shrinkage deformation, which should be determined according to the local experience. If there is no local experience, the value of 0.7 can be used for buildings with three floors or below.

The calculated value of subgrade deformation of buildings on expansive soil subgrade should not be greater than the allowable value of subgrade deformation, that is:

$$s \leqslant [s] \qquad (8\text{-}17)$$

where

s ——the deformation of natural subgrade or treated subgrade, mm;

$[s]$ ——allowable deformation of the subgrade, mm; for expansive soil subgrade, it can be taken according to Table 8-9.

Allowable deformation of expansive soil subgrade   Table 8-9

Structure type	Relative deformation		Deformation/mm
	Type	Value	
Masonry structure	Local tilt	0.001	15
Masonry load-bearing structure with three to four bays and four corners with constructional columns or reinforcement	Local tilt	0.0015	30
Adjacent column subgrades of industrial and civil buildings			
(1) When the frame structure has no infilled wall;	Deformation difference	$0.001l$	30

continued

Structure type	Relative deformation		Deformation/mm
	Type	Value	
(2) When the frame structure has a infilled wall;	Deformation difference	0.0005 l	20
(3) Structure that does not generate additional stress when the subgrade rises and falls unevenly	Deformation difference	0.003l	40

Note: l refers to the center distance between adjacent columns, m.

8.2.4 Engineering measures for expansive soil subgrade

The deformation of expansive soil is extremely sensitive to environmental variations. It is influenced by many external factors and the deformation of expansive soil subgrade is very complicated. In view of the swelling and shrinkage deformation characteristics of expansive soil subgrade, the maximum deformation of subgrade must be strictly controlled not to exceed the allowable deformation. When the deformation requirements are not met, measures should be taken from the aspects of subgrade, foundation, superstructure and construction.

1. Superstructure measures

(1) Buildings should be arranged as far as possible in sites with simple terrain conditions, uniform soil, small terrain gradient and weak swelling and contraction. Buildings should not be built in areas where the groundwater level changes greatly.

(2) The building shape should be simple. Settlement joints should be set at the junction of excavation and fill, or where the subgrade soil is obviously uneven, or at the plane turning part of the building, or the part with significant change in height (load) and different parts of the building structure type.

(3) Waterproof and drainage measures should be strengthen to minimize the water content variation of subgrade soil. Outdoor drainage should be unblocked to avoid water accumulation, and external drainage should be adopted for roof drainage. The apron width should be slightly larger, generally greater than 1.2 m. And thermal insulation layer should be added.

(4) Indoor floor design should be treated differently according to requirements. Measures such as ground reinforcement or overhead should be taken for Grade Ⅲ expansive soil subgrade and the ground with particularly strict use requirements. For general industrial and civil building floors, they can be designed as ordinary floors or paved with precast concrete blocks, but flexible materials should be embedded between the blocks. A large area of the ground should be divided by deformation joints.

(5) Turf should be planted in the open space outside the apron around the building.

When planting trees, we should pay attention to the selection of tree species. For example, we should not plant fast-growing tree species such as eucalyptus, which have the largest water absorption and evaporation. We should choose coniferous trees or shrubs with small evaporation.

2. Structural measures

(1) High-rise buildings with more than three floors should be built in the expansive soil area to increase the basement pressure and prevent swelling deformation.

(2) The strip foundation can be used for the homogeneous weak expansive soil subgrade. The pier foundation should be used if the foundation is buried deeply or the pressure on the strip foundation is small.

(3) The load-bearing masonry structure can adopt solid wall, and the wall thickness should not be less than 240 mm. The bucket wall, block wall or sand-free concrete masonry should not be adopted. Brick arch structure, sand-free macro-porous concrete and unreinforced medium-sized block are not suitable for deformation-sensitive structures.

(4) In order to increase the overall rigidity of the building, ring beams should be set at the top of the subgrade and the top floor of the building. They can interlayer set at other floors of the multi-storey building, or they can also be set layer by layer if necessary.

(5) Steel and reinforced concrete bent structures, gables and internal partition walls should adopt the same foundation as column. The retaining wall should be built on the foundation beam, and there should be a gap of about 100 mm between the bottom of the foundation beam and the ground.

3. Subgrade treatment

The purpose of expansive soil subgrade treatment is to reduce or eliminate the harm caused by subgrade swelling and contraction to buildings. The commonly used methods are given as follows.

(1) Soil replacement cushion

In the construction site where the shallow strong expansive soil layer is exposed, or when the building has strict requirements on uneven deformation, non-expansive cohesive soil, sand and gravel, lime soil, and etc. can be used to replace all or part of the expansive soil, so as to reduce the swelling and shrinkage deformation of the subgrade. The thickness of soil replacement should be determined by deformation requirements. For the Grade I and Grade II expansive soil on flat ground, sand and gravel cushion should be adopted, and the thickness of cushion should not be less than 300 mm. Both sides of subgrade should be backfilled with the same material as cushion. The waterproof and waterproof treatment should be done well.

(2) Increase the buried depth of foundation

For multi-storey buildings on a flat site, when the buried depth of the subgrade is the main prevention measure, the minimum buried depth of the foundation should not be less than the depth of the layer with rapid atmospheric influence.

(3) Lime grouting reinforcement

Adding a certain amount of lime into expansive soil can effectively improve the strength of soil, increase the humidity stability of the soil and reduce the expansion potential. In engineering, lime slurry can be poured into the cracks of expansive soil by pressure grouting to strengthen it.

(4) Pile foundation

When the atmospheric influence is deep and the expansive soil layer is thick, it is difficult or uneconomical to use subgrade reinforcement or pier foundation. In these cases, pile foundation can be used. Pile foundation can also be used for the expansive soil subgrade with swelling and shrinkage Grade Ⅲ or foundation design Grade A. The pile end should be anchored in the non-expansive soil layer or extend into the soil layer below the layer with rapid atmospheric influence. The specific pile foundation design should meet the requirements of *Technical Code for Building in Expansive Soil Region* (GB 50112—2013).

When foundation construction is carried out on expansive soil subgrade, it is advisable to adopt subsection rapid operation method. The subgrade pit shall not be exposed to the sun or soaked in water during the construction process, and waterproof methods shall be taken during the rainy season. After the foundation construction is out of the ground, the subgrade pit shall be backfilled in layers in time.

For sloping site, the expansive soil slope has multi-directional water loss and instability. The buildings on sloping site generally need to be dug and filled, which makes the soil inhomogeneity more prominent. Therefore, the building damage on sloping site is generally more serious than that on flat ground. We should try our best to avoid building houses on this kind of slope. When it is necessary to build a house on the slope, the slope should be treated first and the environment should be improved. After the slope treatment is completed, the building should be started. If the slope is in an unstable state, simple local subgrade treatment is difficult to work. Slope control includes drainage measures, setting retaining and slope protection. The function of slope protection on expansive soil slope is not only to prevent erosion, but also to keep the water content in the slope stable.

Questions

8-1. What are the characteristics of soft soil foundation? How to evaluate soft soil foundation?

8-2 What is the collapsibility of loess? Why does loess have collapsibility? Do all loess have collapsibility?

8-3 What are the main characteristics of collapsible loess?

8-4 What parameters are used to determine the collapsibility of loess?

8-5 What are the engineering measures for collapsible loess subgrade?

8-6 What is expansive soil? What are the main characteristics of the expansive soil?

8-7 How many grades are the expansive soil foundation divided into based on the amount of swelling and shrinkage? How to calculate the amount of swelling and shrinkage deformation?

Chapter 9 Seismic Analysis and Design of Foundations

9.1 Introduction

Strong earthquakes are highly destructive, which often cause a large number of casualties, building damage, traffic and production disruptions, and secondary disasters such as water, fire, and disease. On May 22nd, 1960, Chile experienced the largest earthquake in the world (8.9 on the Richter scale). The disaster was extremely severe. The tsunami caused by the earthquake had a wave height of up to 20 m. When the tsunami crossed the Pacific Ocean and reached the east coast of Japan, the wave height still reached 4-7 m, causing hundreds of casualties and sinking more than 100 ships. This was a serious natural disaster.

China is located between two major seismic belts in the world—the Pacific Rim seismic belt and the Eurasian seismic belt. China experienced several strong earthquakes and serious disaster happened. On July 28th, 1976, an earthquake measuring 7.8 on the Richter scale struck Tangshan, China. The death toll was 242000, and 85% of houses collapsed or were severely damaged. The direct economic losses amounted to tens of billions of yuan. The cost of disaster relief and reconstruction after the earthquake also amounted to nearly 10 billion yuan. The Wenchuan earthquake measuring 8 on the Richter scale on May 12th, 2008 is the strongest earthquake since the founding of P.R. China. The Wenchuan earthquake affected an area of over 500000 km^2, with 87000 people killed and missing and 375000 injured. According to the statistics in September of that year, the direct economic loss was 845.1 billion yuan. We can know how serious the earthquake is from the above two disasters.

As Figure 9-1 shown, various types of ground and building damages can be caused by strong earthquakes due to the complexity of topographical and geological conditions, which will be presented as follows.

1. Ground surface damage

(1) Ground fissures. Ground fissures refer to cracks on the ground subjected to strong earthquakes. Two kinds of ground fissures are usually produced due to different generation reasons: (a) gravity ground fissures. Under the action of earthquakes, the inertial force generated by ground motion exceeds the shear strength of the soil, causing surface cracks, which mostly occur in the deep and saturated soft soil layer of the seashore,

Figure 9-1 Ground and structure damages caused by earthquakes

lakeside, riverbank, slope, and ancient river channel. (b) Structural fractures. A deep fault in the earth's crust that dislocates and extends to the ground during an earthquake.

(2) Sand blasting and water pouring due to soil liquefation. The strong vibration of seismic waves causes the underground saturated fine sand or silt to liquefy. When the underground water pressure increases, the underground water spurts out of the ground through cracks, and together with sand or silt, it spurts out of the ground, forming a phenomenon of sand blasting and water pouring.

(3) Ground collapse. Some extreme earthquake areas generally subside, and some partially subside.

(4) Landslide or collapse of river bank and steep slope.

2. Earthquake damage to engineering facilities

The strong vibration during an earthquake causes damage to various engineering facilities such as buildings, structures, lifeline projects and various equipment.

(1) Damage caused by seismic inertial forces. The structure has insufficient bearing capacity, deformation, or connection strength between structural components. Under the action of seismic inertia forces, the structure cracks, damages, or even collapses.

(2) Damage caused by subgrade failure. When engineering facilities are located on (or in) saturated fine silt, silt, or muddy soft soil, due to seismic liquefaction of the soil or rapid deformation of the muddy soft soil, the bearing capacity of the subgrade decreases, or even completely loses, resulting in damage to the engineering structure, overall dumping, or floating or breaking of the underground structure.

3. Secondary disasters caused by earthquakes

(1) A flood caused by a dam breach.

(2) Fire, explosion, poisoning, pollution, etc. caused by interruption or destruction of lifeline systems (electricity, water, gas, communication, transportation, etc.).

(3) Tsunamis caused by offshore earthquakes.

4. Earthquake damage to foundations

(1) Settlement, uneven settlement, and inclination

The observation data show that the settlement of buildings on general foundations caused by earthquakes is generally small. However, the soft soil foundation can produce a settlement of 10-20 cm, or even more than 30 cm. If the main stress bearing layer of the foundation is liquefiable soil or contains a thick liquefiable soil layer, strong earthquakes may produce settlements of tens of centimeters or even more than 1m, resulting in the tilting and collapse of buildings.

(2) Horizontal displacement

The phenomenon of large horizontal displacement of foundations caused by earthquakes is commonly seen in buildings located on slopes or river banks, due to instability of soil slopes and lateral expansion of underground liquefiable soil layers on the banks.

(3) Tensile failure

During an earthquake, when the outer row of piles of a pile foundation subjected to a large moment of action is subjected to excessive tensile force, the connection between the pile and the pile cap may also suffer tensile damage. The anchor pulling devices of tall structures such as poles and towers may also be damaged due to excessive tensile force generated by earthquakes.

Buildings are built on geotechnical foundations. During the earthquake, the seismic wave propagating in the rock and soil causes the foundation rock and soil to vibrate. Vibration causes additional deformation of soil, and the strength will change accordingly.

When the strength of local foundation soil is greatly reduced under the action of vibration, it will lose its ability to support buildings, leading to foundation failure, and in severe cases, it may cause earthquake damage such as ground fissure, slip, liquefaction and earthquake subsidence. The anti-seismic design of foundation is to study the stability and deformation of foundation in earthquake, including checking the earthquake bearing capacity, judging the liquefaction possibility of foundation, dividing the liquefaction grade, analyzing the earthquake settlement and the anti-seismic measures.

9.2 Site and subgrade

9.2.1 Site classification

Site refers to the land where an engineering group is located and directly used. The same site has similar response spectrum characteristics. The area of one site is equivalent to that of one factory, residential area, village or not less than the area of 1 km^2. The topographic conditions, geology, groundwater level, overburden thickness, and type of site have a great influence on the seismic safety of buildings. The site evaluation is complicated. The *Code for Seismic Design of Buildings* (GB 50011— 2010) (2016 edition) summarizes the experience of earthquake disasters and seismic engineering in China, and refers to many foreign site classification methods, and puts forward the following classification standard. As shown in Table 9-1, according to topography, landforms and geology, it is divided into four sections: favorable, general, unfavorable and dangerous sections for earthquake resistance.

Classification of favorable, general, unfavorable and dangerous sections   Table 9-1

Type	Geology, topography and geomorphology
Favorable section	Stable bedrock, hard soil, open, flat, dense and uniform medium hard soil, etc.
General section	Do not belong to unfavorable, favorable and dangerous areas
Unfavorable section	Soft soil, liquefiable soil, strip-shaped protruding mountain mouth, towering and isolated hills, steep slopes, scarps, the edges of river banks and slopes, soil layers with obviously uneven lithologic state in plane distribution (including old river courses, loose fault fracture zones, buried ponds and valleys and semi-filled and semi-dug foundations), plastic loess with high water content, and structural cracks on the surface, etc.
Dangerous area	Landslides, collapses, subsidence, ground fissures, mudslides, etc. may occur during earthquakes, and parts of the seismogenic fault zone where surface dislocation may occur

Local unfavorable terrain such as strip-shaped protruding shanks, towering isolated hills, steep slopes and scarps may amplify the ground motion parameters. If it is difficult to avoid and buildings must be built in these unfavorable seismic sections, the seismic influence coefficient should be appropriately increased, but it should not be greater than 1.6.

Chapter 9   Seismic Analysis and Design of Foundations

When choosing a site, you should first understand the seismic activity of the site, master the relevant information of engineering geology and seismic geology, and judge the nature of the site. Don't build buildings in dangerous areas, especially Grade A and Grade B buildings. Try to avoid unfavorable areas, and if it is really impossible to avoid them, effective engineering measures should be taken to solve the problems; Strive to build the building on a favorable location.

When there are seismic faults within the site, the engineering impact of the faults should be evaluated. And it should meet the following requirements.

In cases where one of the followingitems is met, the impact of seismic rupture and dislocation on buildings can be ignored:

① The seismic fortification intensity is less than 8-degree;

② Non-Holocene active fault;

③ When the seismic fortification intensity is 8-degree and 9-degree, the overburden thickness of hidden faults is greater than 60 m and 90 m, respectively.

If any of the above items cannot be met, the main fracture zone should be avoided. The dodge distance should not be less than that given in Table 9-2.

Minimum dodge distance for seismic fractures (m) Table 9-2

Seismic intensity	Seismic fortification grade of buildings			
	A	B	C	D
8-degree	Specialized research	200 m	100 m	—
9-degree	Specialized research	400 m	200 m	—

The damage degree of buildings is significantly depended on the soil conditions of the site. The general rule is that compared to hard subgrade, weak subgrade is prone to unstable states, uneven subsidence, and even liquefaction, sliding, cracking, and other phenomena. Meanwhile the seismic damage increases with the overburden thickness. The *Code for Seismic Design of Buildings* (GB 50011—2010) adopts the equivalent shear wave velocity of soil layers and the overburden thickness to classify the site, which is shown in Table 9-3. When there is reliable shear wave velocity and overburden thickness, and their values are near the boundary of the site categories listed in Table 9-3, the interpolation method can be used to determine the design characteristic period for seismic action calculation.

Overburden thickness for various sites (m) Table 9-3

Shear wave velocity of rocks v_s or equivalent shear wave velocity of soil v_{se} (m/s)	Type of site				
	I_0	I_1	II	III	IV
$v_s > 800$	0				
$800 \geqslant v_{se} > 500$		0			

9.2 Site and subgrade

continued

Shear wave velocity of rocks v_s or equivalent shear wave velocity of soil v_{se} (m/s)	Type of site				
	I_0	I_1	II	III	IV
$500 \geqslant v_{se} > 250$		<5	$\geqslant 5$		
$250 \geqslant v_{se} > 150$		<3	3-50	>50	
$v_{se} \leqslant 150$		<3	3-15	15-80	>80

The equivalent shear wave velocity of soil v_{se} can be calculated using the following equations:

$$v_{se} = d_0/t \tag{9-1}$$

$$t = \sum_{i=1}^{n} (d_i/v_{si}) \tag{9-2}$$

where

v_{se} ——equivalent shear wave velocity, m/s;

d_0 ——calculated depth, taking the smaller value of overburden thickness and 20 m, m;

t ——propagation time of shear waves from the ground to the calculated depth, s;

d_i ——thickness of the ith soil layer within the calculated depth, m;

v_{si} ——shear wave velocity of the ith soil layer within the calculated depth, m/s;

n ——number of layers within the calculated depth.

The shear wave velocity of soil should be measured onsite. When there is no measured shear wave velocity, it can be estimated according to the type and properties of the rock and soil, which is shown in Table 9-4.

Classification of soil types and shear wave velocity　　　　Table 9-4

Types of soil	Names and properties	Shear wave velocity /(m/s)
Rock	Hard and intact rock	$v_s > 800$
Hard soil or soft rock	Fractured and relatively fractured rocks or soft rocks, dense gravel soil	$800 \geqslant v_s > 500$
Medium hard soil	Medium-dense, slightly dense gravel soil; dense, medium-dense gravel, coarse sand and medium sand; clay soil and silt with $f_{sk} > 150$kPa; hard loess	$500 \geqslant v_s > 250$
Medium soft soil	Slightly dense gravel, coarse sand and medium sand; fine sand and silty sand with $f_{sk} \leqslant 150$kPa except loose sand; filling soil with $f_{sk} > 130$kPa; new loess can be molded	$250 \geqslant v_s > 150$
Soft soil	Silt and mucky soil; loose sand; recently deposited cohesive soil and silt; filling soil with $f_{sk} \leqslant 130$kPa; flowing loess	$v_s \leqslant 150$

Note: The f_{sk} is the bearing capacity characteristic value of subgrade obtained by field load test, kPa. The v_s is the shear wave velocity of rock and soil.

Chapter 9　Seismic Analysis and Design of Foundations

The overburden thickness should be determined according to the following requirements.

(1) In general, the overburden thickness is the distance from the ground to the top of the soil layer whose shear wave velocity is greater than 500 m/s and all soil layers below has the shear wave velocity no less than 500 m/s.

(2) When there is a soil layer below 5 m from the ground, whose shear wave velocity is greater than 2.5 times the shear wave velocity of each soil layer above it, and the shear wave velocity of its underlying soil layers is not less than 400 m/s, the overburden thickness can be determined based on the distance from the ground to the top surface of the soil layer.

(3) Solitary rocks and lenses with their shear wave velocity greater than 500 m/s should be considered as surrounding soil layers.

(4) The hard interlayer of volcanic rock should be considered as a rigid body, and its thickness should be deducted from the overburden thickness.

9.2.2　Subgrade liquefaction assessment

Loose saturated cohesionless soil and silt tend to shrink in volume under dynamic loading. If the water in the soil cannot be discharged in time, the pore water pressure will increase. When the pore water pressure is accumulated to be equivalent to the overlying pressure of the soil layer, there is no effective pressure between particles, and the soil loses its shear strength. At this time, even if small shear force is applied, viscous flow occurs, which is called liquefaction. Site liquefaction can occur during the process of earthquake or a long time after the earthquake. It often leads to instability, subsidence or excessive uneven settlement of building foundation, which is a serious earthquake damage caused by the earthquake.

Liquefaction assessment is very important for the site. There are various methods to assess the liquefaction possibility of horizontal site (without additional load) or foundation soil. In the following, the method presented in *Code for Seismic Design of Buildings* (*GB 50011—2010*) are described in the following.

When there is saturated sand and saturated silt under the ground, liquefaction assessment should be conducted in two steps except for 6-degree.

1. Preliminary discrimination

Saturated sand or silt (excluding loess) can be preliminarily classified as non-liquefiable soil when one of the following items is met. The impact of liquefaction should not be considered.

(1) When the geologic time scale is Quaternary Late Pleistocene (Q3) and before, it can be classified as non-liquefiable soil for the seismic intensity of 7 and 8 degrees.

(2) When the percentage content of fine particles (particles with their size less than

0.005 mm) in silt is greater than 10, 13, and 16 for the seismic intensity of 7, 8, and 9 degrees, respectively, it can be classified as non-liquefiable soil.

(3) When the thickness of overlying non-liquefiable soil layer and the depth of groundwater level meet one of the following conditions:

$$d_u > d_0 + d_b - 2 \tag{9-3}$$

$$d_w > d_0 + d_b - 3 \tag{9-4}$$

$$d_u + d_w > 1.5d_0 + 2d_b - 4.5 \tag{9-5}$$

where

d_w ——depth of groundwater level, which should be adopted according to the annual average maximum water level during the service life of the building or the maximum water level in the near future, m;

d_u ——thickness of overlying non-liquefiable soil layer, in which silt and muddy soil should be deducted, m;

d ——embedded depth of foundation, and 2 m should be adopted when it is less than 2 m, m;

d_0 ——characteristic depth of liquefiable soil, which can be adopted according to Table 9-5.

Characteristic depth of liquefiable soil/m   Table 9-5

Type of saturated soil	Seismic intensity		
	7-degree	8-degree	9-degree
Silt	6	7	8
Sand	7	8	9

2. Standard penetration test discrimination

When the preliminary identification for saturated sand and silt indicates that the further liquefaction identification should be conducted. As the second step, the standard penetration test discrimination method should be used to determine the liquefaction of soil within 20 m below the ground. But for buildings that do not require seismic bearing capacity verification of natural foundations, only soil within 15 m below the ground should be discriminated. When the standard penetration blow count (without rod length correction) of saturated soil is less than or equal to the critical blow count of standard penetration tests for liquefaction discrimination, it should be judged as liquefiable soil. Otherwise, it should be judged as non-liquefiable soil

The critical blow count of standard penetration tests for liquefaction discrimination can be calculated according to the following equation:

$$N_{cr} = N_0 \beta [\ln(0.6d_s + 1.5) - 0.1d_w] \sqrt{\frac{3}{\rho_c}} \tag{9-6}$$

where

N_{cr} ——critical blow count of standard penetration tests for saturated soil liquefaction;

N_0 ——the reference blow count of standard penetration tests, which can be adopted according to Table 9-6;

β ——adjustment coefficient to consider classification of design earthquake effect: 0.8 for the first group, 0.95 for the second group and 1.05 for the third group;

d_s ——depth of standard penetration testing point of saturated soil, m;

d_w ——depth of groundwater table, m;

ρ_c ——percentage of clay content in saturated soil, when $\rho_c < 3$, take $\rho_c = 3$, %.

Reference blow count of standard penetration tests N_0 Table 9-6

Design basic seismic acceleration /g	0.10	0.15	0.20	0.30	0.40
Reference blow count of standard penetration tests N_0	7	10	12	16	19

The preliminary discrimination and standard penetration test discrimination described above are for a point within a soil column, where there may be multiple points within a soil column. When it is determined that some soil layers in the foundation belong to liquefiable soil, it is necessary to further estimate the liquefaction seriousness of the whole foundation. The liquefaction index I_{lE} is used to evaluate the liquefaction seriousness of the whole foundation, which is calculated using the following equation:

$$I_{lE} = \sum_{i=1}^{n} \left(1 - \frac{N_i}{N_{cri}}\right) h_i W_i \tag{9-7}$$

where

N_i, N_{cri} ——measured standard penetration test blow number and critical blow number of the ith standard penetration testing point in liquefiable soil, when the measured value is greater than the critical value, it is taken as $N_i/N_{cri} = 1.0$;

n ——total number of standard penetration testing points of each borehole within the depth range;

h_i ——the thickness of liquefiable soil layer represented by the ith standard penetration testing point, m;

W_i ——the weight function parameter reflecting the depth of the ith liquefiable soil layer. When the depth of the midpoint of the layer is not greater than 5 m, a value of 10 should be used. When it is equal to 20 m, a value of zero should be used. When it is 5-20 m, a linear interpolation value should be used.

After obtaining the liquefaction index I_{lE}, the subgrade liquefaction grade can be classified according to Table 9-7.

Classification of subgrade liquefaction grade Table 9-7

Liquefaction grade	Slight	Medium	Serious
Liquefaction index I_{lE}	$0 < I_{lE} \leqslant 6$	$6 < I_{lE} \leqslant 18$	$I_{lE} > 18$

9.3 Seismic bearing capacity verification of shallow foundations

9.3.1 Seismic resistance standards

The seismic fortification goal of buildings in China is no damage under small earthquakes, repairable under medium earthquakes and no collapse under large earthquakes. When buildings are subjected to frequent earthquakes lower than the fortification intensity in this area, the main structure of the buildings should have no damage or can be used without repair. When buildings are subjected to the fortification intensity in this area, the building may be damaged to some extent, yet it can still be used after general repair or no repair. When buildings are subjected to seldom occurred earthquakes that are higher than the fortification intensity in this area, the buildings will not collapse or cause serious life-threatening damage.

Of course, the fortification standards of buildings are related to the importance of buildings. According to the importance, buildings are divided into the following four categories.

Grade A buildings are those which have great political, economic and social impacts, or which may cause serious secondary disasters during an earthquake, such as pollution caused by radioactive substances, diffusion of highly toxic gases or big explosions.

Grade B buildings are those whose functions cannot be interrupted during the earthquake or need to be restored as soon as possible, including urban lifeline engineering buildings and buildings needed for disaster relief, such as broadcasting, communication, power supply, water supply, gas supply, ambulance, medical care, firefighting and other buildings.

Grade C buildings are general industrial and civil buildings other than Grade A, Grade B and Grade D buildings.

Grade D buildings are secondary buildings, such as buildings that will not cause casualties and great economic losses when damaged in an earthquake. According to the above design ideas, the fortification standards of various buildings should meet the following requirements.

(1) For Grade A buildings, the seismic action for design should be higher than the re-

quirements of seismic fortification intensity in this area. Its value should be determined according to the approved seismic safety evaluation. The seismic adopted measures should be increased by one degree when the seismic fortification intensity is 6-8 degrees. When the fortification intensity is 9 degrees, it should be higher than the fortification requirement of 9-degree.

(2) For Grade B buildings, the seismic action for design should meet the requirements of seismic fortification intensity in this area. Under normal circumstances, the adopted seismic measures should be increased by one degree according to the intensity of the local area when the fortification intensity is 6-8 degrees, and should be higher than that when it is 9-degree.

(3) For Grade C buildings, the seismic action for design and the seismic measures taken should meet the requirements of seismic fortification intensity in this area.

(4) For Grade D buildings, under normal circumstances, the seismic action for design should still meet the requirements of seismic fortification intensity in this area. The adopted seismic measures are allowed to be lower than the requirements of this area, but they should not be lower when the seismic fortification intensity is 6-degree.

In addition, in the 6-degree fortification area, unless there are special provisions, the calculation of seismic action can be omitted for buildings such as Grade B, Grade C and Grade D. Only seismic measures need to be taken.

9.3.2 Seismic bearing capacity of natural subgrade

Considering that earthquake action is a special working condition, the probability of occurrence is small, the safety of the buildings can be reduced appropriately. The *Code for Seismic Design of Buildings* (GB 50011—2010) required that the seismic bearing capacity of the subgrade should be taken as its bearing capacity characteristic value multiplied by the adjustment coefficient, which is expressed as

$$f_{aE} = \zeta_a f_a \tag{9-8}$$

where

f_{aE} ——seismic bearing capacity of the subgrade soil, kPa;

f_a ——characteristic value of subgrade soil bearing capacity after depth and width correction, which is determined according to the method in Chapter 3;

ζ_a ——adjustment coefficient of seismic bearing capacity of subgrade soil, which can be found in Table 9-8.

Adjustment coefficient of seismic bearing capacity of subgrade soil Table 9-8

Name and properties of subgrade soil	ζ_a
Rock, dense crushed stone, dense gravel, coarse and medium sand, cohesive soil and silt with $f_{ak} \geqslant 300$ kPa	1.5

9.3 Seismic bearing capacity verification of shallow foundations

continued

Name and properties of subgrade soil	ζ_a
Medium-dense and slightly dense crushed stone, medium-dense and slightly dense gravel, coarse sand, dense and medium-dense fine and silty sand, cohesive soil and silt with 150 kPa$\leqslant f_{ak}<$300 kPa, hard loess	1.3
Slightly dense fine and silty sand, cohesive soil and silt with 100kPa$\leqslant f_{ak}<$150kPa, plastic loess	1.1
Silt, muddy soil, loose sand, miscellaneous fill, recently accumulated loess and flowing loess	1.0

When verifying the vertical bearing capacity of natural subgrade under earthquake, the average pressure on the bottom surface of the foundation p_E and the maximum pressure at the edge p_{Emax}, should be calculated according to seismic standard combination. The requirements of Equations 9-9 and 9-10 should be met.

$$p_E \leqslant f_{aE} \tag{9-9}$$

$$p_{Emax} \leqslant 1.2 f_{aE} \tag{9-10}$$

where

p_E ——average pressure on the bottom surface of the foundation under seismic standard combination, kPa;

p_{Emax} ——maximum pressure on the edge of foundation under seismic standard combination, kPa.

In addition, it is required that no tensile stress on the bottom of the foundation is produced under the earthquake for the high-rise building with its height-width ratio greater than 4. That is, the minimum pressure at the edge of the foundation edge p_{Emin} should be no less than 0. For other buildings, the area of zero stress zone between foundation bottom and subgrade surface should not exceed 15% of the total area of foundation bottom.

Example 9-1:

An 8-story building is 25 m high with a raft foundation. The raft foundation has a width of 12 m and a length of 50 m. The subgrade soil is a medium dense fine sand layer. The total vertical force (including the self-weight of the foundation and the above soil) applied on the bottom of the foundation according to the standard combination of seismic load is 100 MN. When the zero-pressure zone of the foundation reaches the maximum limit, try to determine the minimum characteristic bearing capacity of the subgrade soil after depth and width correction in order to meet the seismic bearing capacity requirement.

Solution:

The height-width ratio of the building $= 25/12 = 2.08 < 4$, which indicates that the maximum zero-pressure zone occupies the 15% of the bottom area.

$$3a = (1 - 0.15)b = 0.85b$$

Seismic bearing capacity $f_{aE} = \zeta_a f_a$

The subgrade soil is a medium dense fine sand layer, according to Table 9-8:

$$\zeta_a = 1.3 \Rightarrow f_{aE} = 1.3 f_a$$

Seismic bearing capacity verification:

$$p \leqslant f_{aE} \Rightarrow \frac{N}{A} \leqslant f_{aE} \Rightarrow \frac{100 \times 10^3}{12 \times 50} \leqslant 1.3 f_a \Rightarrow f_a \geqslant 128.2 \text{ kPa}$$

$$p_{max} \leqslant 1.2 f_{aE} \Rightarrow \frac{2N}{3al} \leqslant 1.2 f_{aE} \Rightarrow \frac{2 \times 100 \times 10^3}{0.85 \times 12 \times 50} \leqslant 1.2 \times 1.3 f_a \Rightarrow f_a \geqslant 251.4 \text{ kPa}$$

Take the larger value $f_a \geqslant 251.4$ kPa

9.4 Seismic bearing capacity verification of pile foundations

9.4.1 Damages of pile foundation under earthquake

Based on the investigation on the building damages caused by the 1976 Tangshan earthquake, pile foundation has better seismic performance compared with other types of foundations. Pile foundations play a significant role in reducing the seismic response of the upper structure. However, there are still many seismic damages to pile foundations especially in the 1995 Hanshin earthquake in Japan. The seismic damage of pile foundations is closely related to the characteristics of geological conditions, and the type and load characteristics of the upper structure, as well as the seismic design rationality of pile foundations. Common earthquake damages to piles can be classified into the following types.

(1) The pile head is damaged by shearing, pressing, pulling and bending. Figure 9-2 shows the damage of the pile head of a pile foundation under a high-rise building during the Miyagi earthquake in Japan. Meanwhile it is accompanied by the settlement increase of pile foundation.

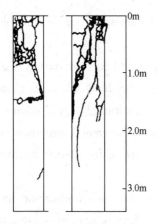

Figure 9-2 Pile head damage during the Miyagi earthquake in Japan

(2) Circumferential cracks are produced when the pile head is bent. It usually occurs in the range of 2-3 m below the pilehead, and the cracks are distributed in the circumferential direction. The reason is that the pile cap has been subjected to excessive bending moment or lateral horizontal force. Figure 9-3 shows the pile foundation failure of Daize Viaduct in Japan. Circumferential cracks generally appear at the top of the pile with its width of 0.5-3 mm.

(3) The pile body is bent due to excessive load on one side of the ground.

Figure 9-4 shows the earthquake damage of the pile foundation in a factory in Tianjin. Because the load on one side reaches 200 kN/m², the bearing capacity of foundation soil is reduced due to earthquake liquefaction. The steel ingot sinks into the soil, laterally squeezing the cast-in-place pile under the pile platform, causing the pile body to break and the pile foundation to tilt.

(4) In the liquefiable soil layer, the pile length does not pass through the liquefiable soil layer, which leads to the failure of the pile foundation. When the foundation is liquefiable soil, if the pile length fails to pass through the liquefiable soil, the pile tip will be supported on viscous liquid with almost no shear

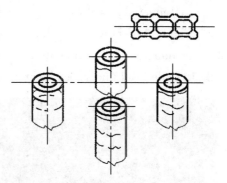

Figure 9-3 Circumferential cracks produced on the pile head

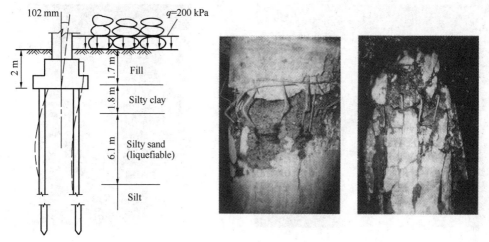

Figure 9-4 Bent damage of pile body

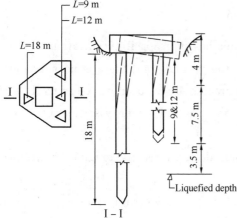

Figure 9-5 Pile damages due to soil liquefaction

strength during liquefaction, which will of course lead to the failure of the pile foundation. Figure 9-5 shows the damage of the pile foundation under the column of a factory in Tianjin during the Tangshan earthquake. There is no backfill around the pile cap, and the liquefaction depth of the foundation is about 15 m. The pile lengths of the two rows of piles are 9 m, 12 m and 18m, and the piles within the liquefaction depth subside, resulting in the long piles being broken due to excessive eccentric pressure.

(5) Liquefiable soil causes pile bending and lateral displacement due to lateral expansion. Figure 9-6 shows the failure of plain concrete piles at the pile head during the Hanshin earthquake. Due to the lateral expansion flow caused by liquefaction of the foundation soil, the pile head undergoes horizontal direct shear failure.

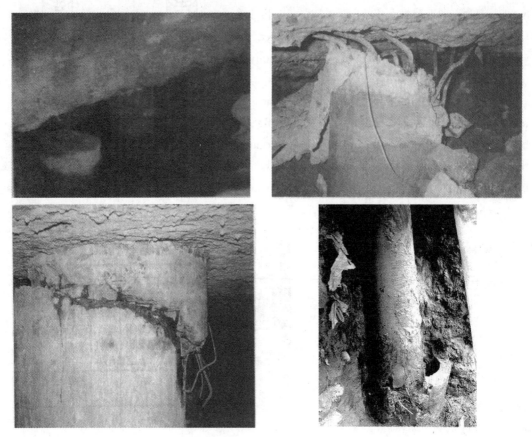

Figure 9-6   Pile damages due to lateral expansion flow of liquefied soil

9.4.2   Seismic bearing capacity of pile foundation

1. Pile foundations that do not require seismic bearing capacity verification

When there is no liquefied soil layer below the ground, and there is no silt, muddy soil, or filling soil with its bearing capacity characteristic value not exceeding 100 kPa around the pile cap, the following buildings are not required to verify seismic bearing capacity for the vertical load pile foundations.

(1) The following buildings with the seismic design intensity ranging from 6-8 degrees:

① General single-story factory buildings and single-story open houses;

② General civil frame houses and frame seismic wall houses with a height of no more than 8 floors and less than 24 m;

③ Multi-story frame factory buildings and multi-story concrete seismic wall buildings with foundation loads equivalent to item ②.

(2) Buildings and masonry buildings meeting the requirements specified in the *Code for Seismic Design of Buildings* (GB 50011—2010).

2. Low capping pile foundation in non-liquefiable soil

The seismic bearing capacity verification of low capping pile foundation in non-liquefiable soil should meet the following requirements.

(1) The characteristic values of vertical and horizontal seismic bearing capacity of a single pile can be increased by 25% compared to non-seismic design.

(2) When the backfill soil around the pile cap is compacted to a density not less than the requirements in the *Code for Design of Building Foundations* (GB 50007—2011), the horizontal seismic effect can be jointly borne by the front fill of the pile cap and the pile. However, the frictional force between the bottom surface of the pile cap and the foundation soil should not be taken into account.

3. Low capping pile foundation in liquefiable soil

The seismic bearing capacity verification of low capping pile foundations with liquefied soil layers should comply with the following regulations:

When the burial depth of the pile cap is shallow, it is not appropriate to consider the resistance of the soil around the pile cap or the sharing effect of the rigid floor on the horizontal seismic action.

When the thickness of non-liquefiable soil or non-weak soil layer is not less than 1.5 m and 1.0 m above and below the foundation bottom respectively, the seismic bearing capacity of the pile can be carried out according to the following two situations and calculated according to one of the unfavorable conditions.

(1) The pile bears all seismic loadings, and the seismic bearing capacity of the pile is increased by 25% compared to non-seismic conditions. The frictional resistance around the pile and the horizontal resistance of the liquefied soil should be multiplied by the reduction factor given in Table 9-9.

(2) The seismic loading is calculated according to 10% of the maximum horizontal seismic impact coefficient, the seismic bearing capacity of the pile is increased by 25% compared to non-seismic conditions, but all frictional resistance of liquefied soil layer and non-liquefied soil within a depth range of 2 m below the pile cap should be deducted.

Reduction factor of soil liquefaction effect Table 9-9

Ratio of measured SPT blow count and critical SPT blow count N/N_{cr}	Depth d_s/m	Reduction factor ψ_L
≤0.6	$d_s \leqslant 10$	0
	$10 < d_s \leqslant 20$	1/3

continued

Ratio of measured SPT blow count and critical SPT blow count N/N_{cr}	Depth d_s/m	Reduction factor ψ_L
>0.6-0.8	$d_s \leqslant 10$	1/3
	$10 < d_s \leqslant 20$	2/3
>0.8-1.0	$d_s \leqslant 10$	2/3
	$10 < d_s \leqslant 20$	1.0

In addition, when the average pile spacing is 2.5-4 times the pile diameter and the number of piles is not less than 5×5, the beneficial effects of driving piles on soil compaction and the deformation limitation of the pile body on liquefied soil can be considered. When the standard penetration blow number of the soil between piles after driving meets the requirement of non-liquefaction, the bearing capacity of a single pile may not be reduced. However, when checking the strength of the bearing layer at the pile tip, the stress diffusion angle outside the pile group should be taken as zero. The standard penetration blow number after pile driving should be determined by testing, and can also be calculated using the following equation:

$$N_l = N + 100\, m\, (1 - e^{-0.3N}) \tag{9-11}$$

where

N_l ——standard penetration blow number after piling;

N ——standard penetration blow number before piling;

m ——area replacement ratio of driven piles.

9.5 Seismic measures for foundations

Earthquake damage of foundation refers to the phenomenon that saturated loose sand or silt in the foundation liquefies, or soft cohesive soil collapses, or the insufficient seismic bearing capacity of the foundation leads to the instability of foundation or the destruction of buildings due to excessive subsidence. Therefore, the engineering measures to improve the bearing capacity of foundation and reduce the deformation and uneven deformation of foundation described in the previous chapters are also effective to improve the seismic capacity of foundation. For example, in the layout of structures, it is required that the building plane and facade should be as regular and symmetrical as possible. The building should have good integrity and stiffness. And the length-height ratio should be controlled in the range of 2-3. Special attention should be paid to the uniform change of lateral stiffness to avoid sudden change. The same structural unit should not be set on the foundation with completely different soil quality. In the layout of the foundation, it is necessary to reasonably increase the buried depth of the foundation and increase the restraining effect of the foundation on the superstructure, so as to reduce the earthquake amplitude of the building, reduce the earthquake damage and increase the overall stability of the foundation. For raft and box foundations of high-rise buildings, the buried depth should not be less than 1/15 of the height of the building. In addition, in the choice of foundation type, we should try to use the foundation with high stiffness and good integrity so as to adjust the additional uneven settlement caused by earthquake. When the foundation is weak cohesive soil, liquefiable soil and newly filled soil, and the soil layer distribution or soil quality is seriously uneven, it is necessary to estimate the uneven settlement or other adverse effects of the foundation caused by the earthquake, and take appropriate foundation reinforcement measures when necessary, such as soil replacement, strong poverty and vibroflotation.

If it is a liquefiable foundation, the anti-liquefaction measures should be taken according to the liquefaction grade of the foundation and the type of buildings, as shown in Table 9-10.

Anti-liquefaction measures   Table 9-10

Seismic fortification type of buildings	Liquefaction grade of foundation		
	Light	Medium	Serious
Grade B	Partial elimination of liquefaction settlement; or foundation and superstructure treatment	Completely eliminate liquefaction settlement or partially eliminate liquefaction settlement; and foundation and superstructure treatment	Completely eliminate liquefaction settlement
Grade C	For the treatment of foundation and superstructure, no measures may be taken	Foundation and superstructure treatment, or measures with higher requirements	Completely eliminate liquefaction settlement or partially eliminate liquefaction settlement; and foundation and superstructure treatment
Grade D	No measures should be taken	No measures should be taken	Foundation and superstructure treatment, or other economic measures

Note: Anti-liquefaction measures for Class A buildings should have special research, which should not be lower than that for Class B buildings.

(1) The measures to completely eliminate liquefaction settlement of the foundation should meet the following requirements.

① When using a pile foundation, the length of the pile end extending into the stable soil layer below the liquefaction depth (excluding the pile tip) should be determined by calculation. It should not be less than 0.8 m for crushed stone, gravel, coarse, medium sand, hard cohesive soil, and dense silt. And it should not be less than 1.5 m for other soils.

② When using a deep foundation, the bottom surface of the foundation should be buried in a stable soil layer below the liquefaction depth with its depth not be less than 0.5 m.

③ When using densification methods (such as vibro-flotation, vibration densification, crushed stone pile compaction, strong labor, etc.) for reinforcement, it should be treated to the limit liquefaction depth. After placing vibro-flotation or compaction gravel piles to soils, the standard penetration blow number of the soil should not be less than the critical value for the liquefaction discrimination criteria.

④ Replace all liquefiable soil layers with non-liquefiable soil or increase the thickness of the overlying non-liquefiable soil layer.

⑤ When using compaction or replacement methods for improving the soil, the improvement width outside the foundation edge should exceed 1/2 of the treatment depth below the foundation bottom and not be less than 1/5 of the foundation width.

(2) Measures to partially eliminate liquefaction settlement of the foundation should meet the following requirements.

① The treatment depth should reduce the liquefaction index of the treated foundation, and its value should not be greater than 5. The liquefaction index of the center area of large raft and box foundations after treatment can be reduced by 1.0 compared to the above regulations. For pad foundations and strip foundations, they should not be less than the larger values of the characteristic depth and foundation width of liquefaction under the foundation bottom surface.

② After using vibro-flotation or compaction gravel piles for reinforcement, the standard penetration blow number should not be less than the critical value.

③ The treatment width beyond the edge of the foundation should comply with the requirements specified in the *Code for Seismic Design of Buildings* (GB 50011—2010).

④ Adopt other methods to reduce liquefaction induced settlement, such as increasing the thickness of the overlying non-liquefiable layer and improving the surrounding drainage conditions.

(3) Measures taken to mitigate the impact of liquefaction on the foundation and upper structure.

① Choose the appropriate buried depth of foundation.

② Adjust the bottom area of the foundation to reduce foundation eccentricity.

③ Strengthen the integrity and stiffness of the foundation, such as using box foundations, raft foundations, or reinforced concrete cross strip foundations, and adding foundation ring beams.

④ Reduce load, enhance the overall stiffness and uniform symmetry of the upper structure, reasonably set settlement joints, and avoid using structural forms sensitive to uneven settlement.

⑤ At the point where the pipeline passes through the building, sufficient size should be reserved or flexible joints should be used.

Questions

9-1 What is the impact of sites on earthquake action? What are the influencing factors?

9-2 Please briefly explain the mechanism of soil liquefaction. Why is saturated loose sand prone to liquefaction and saturated dense sand less prone to liquefaction?

9-3 How to preliminarily determine whether the subgrade soil layer may liquefy?

9-4 How to determine whether the subgrade soil layer can be liquefied based on the results of standard penetration tests? What criteria are used to determine?

9-5 Should the seismic bearing capacity of the subgrade be reduced or increased considering seismic effects? Why?

9-6 How to determine the bearing capacity of piles when there is a liquefied soil layer in the subgrade soil layers?

9-7 The pile foundation has good seismic performance. Can you explain the reason?

Chapter 10　Comparison for Foundation Engineering Design Codes

10.1　Introduction to design codes in different countries

The design of foundation engineering is different from that of other general civil engineering structures. It includes the interaction between subgrade soil and foundations. And it is necessary to obtain the engineering characteristics of subgrade soil as much as possible. Therefore, the design codes in different countries mainly includes two aspects. One is the subgrade soil investigation, which includes various in-situ tests, laboratory tests, and etc. The other is the design of related engineering structures based on the engineering characteristics of soils.

The design codes in different countries are different. In China, due to different industry divisions, domestic regulations in China are divided into several major industry regulations, including those of the Ministry of Housing and Urban-rural Development, the Ministry of Transportation, and the Ministry of Water Resources and Hydropower. These regulatory systems within different industries overlap and have different focuses. With the accumulation and development of engineering experience, these standards gradually tend to be consistent.

At present, national standards such as *Code for Investigation of Geotechnical Engineering (GB 50021—2001)*, *Standard for Geotechnical Testing Method (GB/T 50123—2019)*, *Standard for Test Methods of Engineering Rock Mass (GB/T 50266—2013)*, and *Standard for Engineering Classification of Rock Mass (GB/T 50218—2014)* are commonly used in several major industries in China. For the design of shallow foundations, pile foundations, retaining structures and ground improvement, the national codes include the *Code for Design of Building Foundation (GB 50007—2011)*, *Technical Code for Building Pile Foundations (JGJ 94—2008)*, *Technical Specification for Retaining and Protection of Building Foundation Excavations (JGJ 120—2012)*, *Technical Code for Ground Treatment of Buildings (JGJ 79—2012)*, and etc. In addition, there are many technical regulations, manuals, and local codes to supplement these specifications.

The obvious feature of the above codes is that the content of each code is relatively detailed and independent. For example, designers can basically complete the design calculation of pile foundation according to *Technical Code for Building Pile Foundations (JGJ*

94—2008). And part of its content overlaps with the *Code for Design of Building Foundation (GB 50007—2011)*. Industry specifications focus on the characteristics of the industry. For example, *Code for Design of Building Foundation (GB 50007—2011)* and *Specification for Design of Foundation of Highway Bridge and Culverts (JTG 3363—2019)* are applicable to similar geotechnical structures in the construction industry and transportation industry respectively. These two specifications have strong similarity in content, yet each has obvious industry characteristics. The introduction to design codes in Europe and the USA will be detailed as follows.

10.1.1 Introduction to Eurocodes

The current Structural Eurocodes has been prepared by Technical Committee CEN/TC250, the secretariat of which is held by BSI. CEN/TC250 is responsible for all Structural Eurocodes. In 1975, the Commission of the European Community decided on an action program in the field of construction, based on article 95 of the Treaty. The objective of the program was to establish a set of harmonized technical rules for the design of construction works and eliminate technical obstacles to trade.

Table 10-1 shows the components of the Structural Eurocode. It totally comprises 10 standards, and each consists of a number of Parts. Among them, Eurocode 7 is the European Code for Geotechnical Design. It contains two parts: Part 1 is basic rules and Part 2 is ground investment and testing. The detailed content and arrangement in the Part 1 are given in Table 10-2. It includes twelve main subjects and nine appendices. Eurocode 7 mainly focuses on design principles and the main factors to be considered in design, and relatively few specific calculation methods are mentioned in the main subjects. The calculation formulas and charts provided in the appendices are for reference and are not mandatory by regulations.

Components of Structural Eurocode Table 10-1

Number	Code name	Title	Number of parts
EN 1990	Eurocode	Basis of structural design	1
EN 1991	Eurocode 1	Actions on structures	40
EN 1992	Eurocode 2	Design of concrete structures	4
EN 1993	Eurocode 3	Design of steel structures	20
EN 1994	Eurocode 4	Design of composite steel and concrete structures	3
EN 1995	Eurocode 5	Design of timber structures	3
EN 1996	Eurocode 6	Design of masonry structures	4
EN 1997	Eurocode 7	Geotechnical design	2
EN 1998	Eurocode 8	Design of structures for earthquake resistance	6
EN 1999	Eurocode 9	Design of aluminium structures	5

Chapter 10 Comparison for Foundation Engineering Design Codes

Contents of Eurocode 7: Geotechnical design-Part 1: General rules   Table 10-2

Section number	Title
Section 1	General
Section 2	Basis of Geotechnical Design
Section 3	Geotechnical Data
Section 4	Supervision of Construction, Monitoring and Maintenance
Section 5	Fill, Dewatering, Ground Improvement and Reinforcement
Section 6	Spread Foundations
Section 7	Pile Foundations
Section 8	Anchorages
Section 9	Retaining Structures
Section 10	Hydraulic failure
Section 11	Overall stability
Section 12	Embankments
Annex A	Partial and correlation factors for ULS states and recommended values
Annex B	Background information on partial factors for Design Approaches 1,2 and 3
Annex C	Sample procedures to determine limit values of earth pressures on vertical walls
Annex D	A sample analytical method for bearing resistance calculation
Annex E	A sample semi-empirical method for bearing resistance estimation
Annex F	Sample methods for settlement evaluation
Annex G	A sample method for deriving presumed bearing resistance for spread foundations on rock
Annex H	Limiting values of structural deformation and foundation movement
Annex I	Checklist for construction supervision and performance monitoring

According to the complexity and vulnerability of geotechnical engineering, Eurocode 7 categorizes geotechnical engineering risk into three levels: low, medium, and high. And then Eurocode 7 classified the geotechnical engineering into problems three categories: GC1, GC2, and GC3. Based on this classification, the requirements for personnel, survey, engineering design process, and etc. required for different types of geotechnical engineering problems are proposed. However, this classification is not a mandatory provision and it is optional.

The main design concept of Eurocode 7 is limit state design theory, which is clearly stated in the EN1990 structural design foundation. The specification requires a clear distinction between the ultimate limit state of bearing capacity (ULS) and serviceability limit State of the normal function (SLS). Different calculations are used to verify ULS and SLS.

The characteristics of Eurocode are given as follows. Firstly, it is based on the large civil engineering industry and can be widely applied in civil engineering. Secondly, the generalization of norms in Eurocode is relatively high due to many member countries Finally, it widely refers to foreign codes and classic academic works.

10.1.2　Introduction to USA Standards

The main USA Standards on structural design include International Building Code (IBC), Building Code Requirements for Structural Concreter (ACI318M-05), Specification for Structural Steel Buildings (ANSI/AISC 360-05), National Design Specification for Wood Construction (ANSI/AF&PA NDS-2005), AASHTO LRFD 2012 Bridge Design Specifications 6th Ed (US), and etc.

IBC is written by the International Code Council (ICC) in the United States. The ICC was established in 1994 with the aim of establishing a complete set of unified, comprehensively coordinated, and domestically highest-level code (Model Code). The establishment of a unified code has many advantages. Firstly, throughout the United States, all of the owners, designers and contractors face consistent requirements. Secondly, manufacturers can conduct more new research and development, rather than adapting to different regulatory requirements. Thirdly, unified education and certification can be achieved throughout the country, promoting domestic and international exchanges. Finally, the specification writing departments in various states and regions can fully adopt the specifications written by ICC, thereby putting more effort into expanding the application of the specifications and better implementing them.

IBC consists of 35 chapters and 10 appendices, which covers all aspects of structures including management, design, materials, and etc. Chapter 16 is about structural design, which mainly includes basic structural design requirements, load combinations, and the values of various loads. The following chapters are about various building materials such as concrete, steel, and wood, and etc. Chapter 18 focuses on soil and foundation.

The AASHTO LRFD 2012 Bridge Design Specifications 6th Ed (US) is issued by the Transportation Authority (AASHTO). This specification covers a wide range of content, including all aspects of bridge design. Chapter 4 involves the analysis and calculation of retaining structures, and Chapter 10 mainly discusses foundation design. Other design calculations of geotechnical structures are scattered in various chapters. The main content of this specification is the design of various geotechnical structures. The investigation mainly comes from other codes of AASHTO and ASTM standards.

10.2　Comparison of the bearing capacity verification for shallow foundations

10.2.1　Comparison of bearing pressure calculation

To calculate the bearing pressure on the bottom of the shallow foundations, the equivalent base stress method is generally used in Chinese code of *Code for Design of Building Foundation* (*GB 50007—2011*) when foundations are subjected to eccentric loads. Depending on the eccentricity, the base bearing pressure is generally distributed in a triangular or trapezoidal shape, which is described in Chapter 3 in detail.

Eurocode and AASHTO LRFD 2012 Bridge Design Specifications 6th Ed (US) adopt the equivalent contact area method to calculate the bearing pressure on the bottom of the shallow foundations. It assumes that the stress on the base is evenly distributed. However, the effective contact area decreases due to the influence of eccentric loads. Figure 10-1 shows the reduced dimensions for a rectangular footing. The reduced dimensions for an eccentrically loaded rectangular footing are taken as Equation 10-1 and Equation 10-2. The bearing pressure is calculated according to Equation 10-3. Equation 10-4 should be met for the bearing capacity verification.

$$B' = B - 2e_B \tag{10-1}$$

$$L' = L - 2e_L \tag{10-2}$$

$$p = \frac{R}{B' \times L'} \tag{10-3}$$

where

B——width of the footing, m;

L——length of the footing, m;

e_B ——eccentricity parallel to dimension B, m;

e_L ——eccentricity parallel to dimension L, m;

R——vertical loads, kN.

$$p \leqslant f_a \tag{10-4}$$

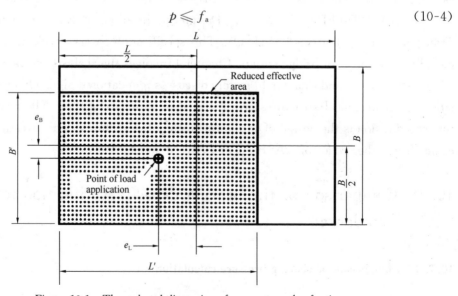

Figure 10-1 The reduced dimensions for a rectangular footing

10.2 Comparison of the bearing capacity verification for shallow foundations

where

f_a ——bearing capacity of the subgrade soil under the bottom surface of the footing, kPa.

10.2.2 Comparison of bearing capacity calculation

There are generally two ways to determine the bearing capacity of a foundation. One is the ultimate bearing capacity theory based on the shear strength parameter of the subgrade soil, the other is determined based on on-site test data combined with engineering experience. As far as the ultimate bearing capacity theory is considered, the Chinese, US, and European specifications are based on the Terzaghi ultimate bearing capacity theory and its developed theories. The method to calculate bearing capacity in Chinese code of *Code for Design of Building Foundation* (GB 50007—2011) is described in Chapter 3 in detail. In the following, the method to calculate bearing capacity in US code is presented.

The bearing capacity of a soil layer is taken as:

$$q_n = cN_{cm} + \gamma D_f N_{qm} C_{wq} + 0.5\gamma B N_{\gamma m} C_{w\gamma} \tag{10-5}$$

in which:

$$N_{cm} = N_c s_c i_c \tag{10-6}$$
$$N_{qm} = N_q s_q d_q i_q \tag{10-7}$$
$$N_\gamma m = N_\gamma s_\gamma i_\gamma \tag{10-8}$$

where

c ——cohesion, taken as undrained shear strength, ksf;

N_c ——cohesion term (undrained loading) bearing capacity factor as specified in Table 10-3;

N_q ——surcharge (embedment) term (drained or undrained loading) bearing capacity factor as specified in Table 10-3;

N_γ ——unit weight (footing width) term (drained loading) bearing capacity factor as specified in Table 10-3;

γ ——total (moist) unit weight of soil above or below the bearing depth of the footing, kcf;

D_f ——footing embedment depth, ft;

B ——footing width, ft;

$C_{wq}, C_{w\gamma}$ ——correction factors to account for the location of the groundwater table as specified in Table 10-4;

s_c, s_γ, s_q ——footing shape correction factors as specified in Table 10-5;

d_q ——correction factor to account for the shearing resistance along the failure surface passing through cohesionless material above the bearing elevation as specified in Table 10-6;

i_c, i_γ, i_q ——load inclination factors determined from Equations 10-9 or 10-10, and Equations 10-11 and 10-12.

For $\phi_f=0$,
$$i_c = 1-(nH/cBLN_c) \tag{10-9}$$

For $\phi_f>0$,
$$i_c = i_q-[(1-i_q)/(N_q-1)] \tag{10-10}$$

in which
$$i_q = \left(1-\frac{H}{V+cBL\cot\phi_f}\right)^n \tag{10-11}$$

$$i_\gamma = \left(1-\frac{H}{V+cBL\cot\phi_f}\right)^{(n+1)} \tag{10-12}$$

$$n = [(2+L/B)/(1+L/B)]\cos^2\theta + [(2+B/L)/(1+B/L)]\sin^2\theta \tag{10-13}$$

where

B ——footing width, ft;

L ——footing length, ft;

H ——unfactored horizontal load, kips;

V ——unfactored vertical load, kips;

θ ——projected direction of load in the plane of the footing, measured from the side of length L (degrees), as Figure 10-2 shown.

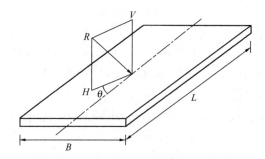

Figure 10-2 Inclined loading conventions

Bearing capacity factors N_c, N_q, and N_γ Table 10-3

ϕ_f	N_c	N_q	N_γ	ϕ_f	N_c	N_q	N_γ
0	5.14	1.0	0.0	10	8.4	2.5	1.2
1	5.4	1.1	0.1	11	8.8	2.7	1.4
2	5.6	1.2	0.2	12	9.3	3.0	1.7
3	5.9	1.3	0.2	13	9.8	3.3	2.0
4	6.2	1.4	0.3	14	10.4	3.6	2.3
5	6.5	1.6	0.5	15	11.0	3.9	2.7
6	6.8	1.7	0.6	16	11.6	4.3	3.1
7	7.2	1.9	0.7	17	12.3	4.8	3.5
8	7.5	2.1	0.9	18	13.1	5.3	4.1
9	7.9	2.3	1.0	19	13.9	5.8	4.7

10.2 Comparison of the bearing capacity verification for shallow foundations

continued

ϕ_f	N_c	N_q	N_γ	ϕ_f	N_c	N_q	N_γ
20	14.8	6.4	5.4	33	38.6	26.1	35.2
21	15.8	7.1	6.2	34	42.2	29.4	41.1
22	16.9	7.8	7.1	35	46.1	33.3	48.0
23	18.1	8.7	8.2	36	50.6	37.8	56.3
24	19.3	9.6	9.4	37	55.6	42.9	66.2
25	20.7	10.7	10.9	38	61.4	48.9	78.0
26	22.3	11.9	12.5	39	67.9	56.0	92.3
27	23.9	13.2	14.5	40	75.3	64.2	109.4
28	25.8	14.7	16.7	41	83.9	73.9	130.2
29	27.9	16.4	19.3	42	93.7	85.4	155.6
30	30.1	18.4	22.4	43	105.1	99.0	186.5
31	32.7	20.6	26.0	44	118.4	115.3	224.6
32	35.5	23.2	30.2	45	133.9	134.9	271.8

Coefficients C_{wq} and $C_{w\gamma}$ for Various Groundwater Depths — Table 10-4

D_w	C_{wq}	$C_{w\gamma}$
0.0	0.5	0.5
D_f	1.0	0.5
$> 1.5B + D_f$	1.0	1.0

Shape correction factors s_c, s_γ, s_q — Table 10-5

Factor	Friction angle	Cohesion Term (s_c)	Unit Weight Term (s_γ)	Surcharge Term (s_q)
Shape factors s_c, s_γ, s_q	$\phi_f = 0$	$1 + \left(\dfrac{B}{5L}\right)$	1.0	1.0
	$\phi_f > 0$	$1 + \left(\dfrac{B}{5L}\right)\left(\dfrac{N_q}{N_c}\right)$	$1 - 0.4\left(\dfrac{B}{L}\right)$	$1 + \left(\dfrac{B}{L}\tan\phi_f\right)$

Depth correction factor d_q — Table 10-6

Friction Angle, ϕ_f (°)	D_f/B	d_q
32	1	1.20
	2	1.30
	4	1.35
	8	1.40
37	1	1.20
	2	1.25
	4	1.30
	8	1.35

continued

Friction Angle, ϕ_f (°)	D_f/B	d_q
42	1	1.15
	2	1.20
	4	1.25
	8	1.30

As far as the bearing capacity verification of shallow foundations is considered, the bearing capacity determined by the Chinese code is generally smaller than the design value determined by European and US codes. At the same time, when determining the base bearing pressure under eccentric load, Chinese code considers the stress redistribution caused by uniformly distributed load and eccentric load and a linear distribution of base bearing pressure is produced. However, the European and US codes believe that the eccentric load causes a decrease in the effective contact area between the foundation and the subgrade soil and the load is evenly borne by the effective contact area. As a result, the base bearing pressure calculated according to European and US codes is greater than the average base bearing pressure and less than the maximum base bearing pressure determined in Chinese code. When the internal friction angle of the subgrade soil is relatively large, the bearing capacity of the foundation determined by the European and US codes is about 30% greater than Chinese code. The design safety coefficient designed by the European and US codes is about 25% greater than that of Chinese code.

10.3 Comparison of vertical bearing capacity calculation of piles

After comparing the calculation methods of bearing capacity of piles in Chinese, US and European codes, it is found that there are a lot in common. For example, the bearing capacity of piles is composed of side friction and tip resistance. The pile side friction can be obtained through the unit friction and area between the soil and the pile. The significant difference in the calculation of the pile bearing capacity mainly lies in how to take the value of the unit friction between the pile and soil and the tip resistance. In the following, the methods of static load tests, CPT tests, SPT tests and PMT tests are compared for Chinese, US and European codes.

1. Static load test

Static load test is the most reliable test method for determining the bearing capacity of pile foundations. All the Chinese, US, and European codes unanimously adopt static load tests as the direct method to determine the bearing capacity of pile foundations. And the static load test is also used to verify pile design.

There is a lot of consistency in the Chinese and European codes when determining the pile bearing capacity through static load tests. When the pile static load test shows a failure state, the previous load level is used as the maximum bearing capacity. The failure state can be determined by drawing Q-s curve, s-lgt curve, or determined based on on-site tests. When there is no obvious failure state, the load corresponding to the specified settlement amount is taken as the ultimate load. The specified settlement amount takes 5% of the pile diameter in Chinese code and 10% of the pile diameter in European code.

In the AASHTO LRFD 2012 Bridge Design Specifications (6th Ed) (US), a static pile load test follows the procedures specified in ASTM D1143, and the loading procedure follows the Quick Load Test Procedure. The pile bearing resistance is determined from the test data as follows:

(1) For piles 24 in (610 mm) or less in diameter (length of side for square piles), the Davisson Method;

(2) For piles larger than 36 in (914 mm) in diameter (length of side for square piles), at a pile top movement, s_f(in), as determined from Equations 10-4;

(3) For piles greater than 24 in (610 mm) but less than 36 in (914 mm) in diameter, criteria to determine the pile bearing resistance is linearly interpolated between the criteria determined at diameters of 24 and 36 in.

$$s_f = \frac{QL}{12AE} + \frac{B}{2.5} \qquad (10\text{-}14)$$

where

Q——test load, kips;

L——pile length, ft;

A——pile cross-sectional area, ft^2;

E——pile modulus, ksi;

B——pile diameter (length of side for square piles), ft.

2. Cone penetration test (CPT)

The CPT is applicable to almost all types of soil. All the Chinese, US, and European codes take CPT as an important method to determine the pile bearing capacity. And various empirical formulas are established for the end resistance, side friction resistance. The CPT is divided into single bridge static penetration test and double bridge static penetration test. Currently, the double bridge static penetration test is widely used, which can provide side wall friction resistance and cone tip resistance.

There is similarity in the application of CPT results in Chinese and US codes. CPT is used to determine the side resistance and the tip resistance of piles. Equation 10-15 is used to calculate the bearing capacity in Chinese code. In US code, Equations 10-16 and 10-17 are used to calculate the side resistance and the tip resistance of piles respectively. By comparing Equations 10-15 to 10-17, there are also some differences between Chinese and US codes. In US code, for the soil layer with the depth less than 8D (D is the pile diameter), reduction is required for the layer, and the reduction coefficient is the ratio of the depth to 8D.

$$Q_{uk} = Q_{sk} + Q_{pk} = u \Sigma l_i \cdot \beta_i \cdot f_{si} + \alpha \cdot q_c \cdot A_p \qquad (10\text{-}15)$$

where

β_i ——correction coefficient, for cohesive soil and silt $\beta_i = 10.04\ (f_{si})^{-0.55}$, for sand $\beta_i = 5.05\ (f_{si})^{-0.45}$;

f_{si} ——average side resistance in the ith layer of soil, kPa;

α ——correction coefficient for pile tip resistance, which takes 2/3 for cohesive soil and silt and 1/2 for saturated sand;

q_c ——The resistance above and below the pile end plane, which is taken as the weighted average value of the resistance within the range of 4d above the pile end plane and 1d below the pile end plane (d is the diameter or side length of the pile).

$$R_s = K_{s,c} \left[\sum_{i=1}^{N_1} \left(\frac{L_i}{8D_i} \right) f_{si} a_{si} h_i + \sum_{i=1}^{N_2} f_{si} a_{si} h_i \right] \qquad (10\text{-}16)$$

where

$K_{s,c}$ ——correction factors, K_c for clays and K_s for sands from Figure 10-3;

L_i ——depth to middle of length interval at the point considered, ft;

D_i ——pile width or diameter at the point considered, ft;

f_{si} ——unit local sleeve friction resistance from CPT at the point considered, ksf;

a_{si} ——pile perimeter at the point considered, ft;

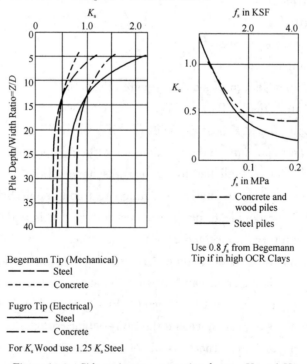

Figure 10-3 Side resistance correction factors K_s and K_c

h_i ——length interval at the point considered, ft;

N_1 ——number of intervals between the ground surface and a point $8D$ below the ground surface;

N_2 ——number of intervals between $8D$ below the ground surface and the tip of the pile.

$$q_p = \frac{q_{c1} + q_{c2}}{2} \quad (10\text{-}17)$$

where

q_{c1} ——average q_c over a distance of yD below the pile tip (path a-b-c); sum q_c values in both the downward (path a-b) and upward (path b-c) directions; use actual q_c values along path a-b and the minimum path rule along path b-c; compute q_{c1} for y-values from 0.7 to 4.0 and use the minimum q_{c1} value obtained (ksf), which can be seen in Figure 10-4;

q_{c2} ——average q_c over a distance of $8D$ above the pile tip (path c-e); use the minimum path rule as for path b-c in the q_{c1}, computations; ignore any minor "x" peak depressions if in sand but include in minimum path if in clay (ksf), which can be seen in Figure 10-4.

Compared to Chinese and US codes, The European code is different. It uses the cone tip resistance to obtain the side friction, which is given in the following:

$$P_{\max;\text{shaft};z} = a_s \cdot q_{c;z;a} \quad (10\text{-}18)$$

where

$P_{\max;\text{shaft};z}$ ——the pile side frictional resistance at a certain depth;

a_s ——conversion coefficient;

$q_{c;z;a}$ ——cone tip resistance at this depth.

3. Standard penetration test (SPT)

The standard penetration test (SPT) has been widely applied in China, US, and Europe. Various empirical formulas have been proposed. In recent years, China has also carried out SPT on cohesive soil and established some empirical relationship. However, the same as in Europe and the United States, the SPT is mostly used for sandy soil to determine the bearing capacity of piles. It must be combined with other methods to complete the design of pile foundation.

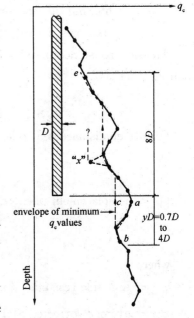

Figure 10-4 Pile end-bearing computation procedure

In the application of SPT, Chinese codes use standard penetration blow count N or the standard penetration blow count N_1 after rod length correction. However, in Europe and US, the standard penetration blow count N_{60}, which has been corrected for energy efficiency, is mainly used. Professional instruments are required to measure the stress waves transmitted to the penetrator and calculate energy efficiency. Therefore, it has higher requirements for the testing instruments and personnel of the SPT. By using SPT, N_{60} is more rigorous in Europe and America, and the empirical formula obtained is more valuable for wide application. However, it requires precise measuring instruments and the post processing is relatively complex. In Chinese codes, the application of SPT is much simpler. The SPT method of the US code to determine bearing capacity of piles is given below.

The unit tip resistance for piles driven to a depth D_b into a cohesionless soil stratum shall be taken as:

$$q_p = \frac{0.8(N1_{60})D_b}{D} \times q_l \qquad (10\text{-}19)$$

where

$N1_{60}$ ——representative SPT blow count near the pile tip corrected for overburden pressure, blows/ft;

D——pile width or diameter, ft;

D_b——depth of penetration in bearing strata, ft;

q_l ——limiting tip resistance taken as eight times the value of $N1_{60}$ for sands and six times the value of $N1_{60}$ for nonplastic silt, ksf.

The side resistance of piles in cohesionless soils takes the Meyerhof method and it shall be taken as:

$$q_p = \frac{0.8(N1_{60})D_b}{D} \times q_l \qquad (10\text{-}20)$$

For driven displacement piles,

$$q_s = \frac{\overline{N1}_{60}}{25} \qquad (10\text{-}21)$$

For nondisplacement piles, e. g., steel H-piles,

$$q_s = \frac{\overline{N1}_{60}}{50} \qquad (10\text{-}22)$$

where

q_s ——unit side resistance for driven piles, ksf;

$\overline{N1}_{60}$ ——average corrected SPT-blow count along the pile side, blows/ft.

4. Pressuremeter test (PMT)

The PMT is suitable for various types of soil and soft rock, and is a widely used in-situ testing method. In Chinese, US, and European codes, the test details and application scope of PMT are described, but it is rarely used for pile design in Chinese and US codes.

Although the description of PMT is detailed and complete in Chinese codes, the PMT

10.3 Comparison of vertical bearing capacity calculation of piles

is not widely used in China and there is no corresponding empirical formulas for the PMT to determine the bearing capacity of piles. The relevant empirical formulas are not provided in the American AASHTO specifications.

The PMT are widely used in the Europe, and empirical formulas have been proposed to determine the bearing capacity of piles:

$$Q = A \times k \times [P_{LM} - P_0] + u \Sigma [q_{si} \times z_i] \tag{10-23}$$

where

A ——the cross-sectional area of the pile end;

k ——the conversion coefficient;

P_{LM} ——the ultimate pressure;

P_0 ——the self weight stress;

u ——the circumference of the pile;

z_i ——the thickness of the formation;

q_{si} ——the lateral frictional resistance of the pile determined by the PMT.

The PMT has a special advantage. The parameters of the PMT can be used to estimate the design parameters of all strata for the pile bearing capacity calculation. The PMT doesn't need other tests as supplements.

Example 10-1:

A static load test was conducted on the piles of a bridge project, and a double bridge static penetration test was used to measure the side wall friction resistance and cone tip resistance. The Osterberg method was used for the pile static load test, and two piles were tested. The test pile SZ1 had a diameter of 1500 mm and a length of 76.1 m. The test pile SZ2 has a diameter of 1200 mm and a length of 60.0 m. Both of them are friction piles.

The engineering geological data is given as follows: ① Plain fill with a thickness of 0-4.2 m; ②^1 Silty clay with a thickness of 0-1.5 m; ②^2 Silty clay with a thickness of 0-5.0 m; ③ Clay with a thickness 0-3.5 m; ④ Silty clay and silt interbedded with a thickness of 1.5-5.4 m; ⑤ Silty soil mixed with silt with a layer thickness of 3.0-7.7 m; ⑥ Clay with a thickness of 2.7-6.8 m; ⑦ Silt with a layer thickness of 0-6.7 m; ⑧ Silty clay and silt interbedded with a layer thickness of 4.5-15.0 m; ⑨ Clay with a layer thickness of 0-9.2 m; ⑩ Silty clay and silt interbedded with a thickness of 1.5-9.6 m; ⑩^A Silt mixed with silt with a layer thickness of 0-9.9 m; ⑪ Clay with a layer thickness of 1.2-10.0 m; ⑪^A Silty clay and silt interbedded with a layer thickness of 0~3.1 m; ⑫ A Gravel soil with a layer thickness of 1.0-3.6 m; ⑬^1 Completely weathered tuffaceous sandstone with a layer thickness of 0-10.8 m; ⑬^2 Moderately weathered tuffaceous sandstone with a layer thickness of 0-9.0 m; ⑭ Moderately weathered angular gravel with a layer thickness of 0-6.6 m; ⑮^1 Strongly weathered silty mudstone, missing at the test pile; ⑮^2 Moderately weathered silty mudstone. Please compare the bearing capacity of piles using different codes.

Solution:

Table 10-7 gives the bearing capacity values measured in the pile static load tests The bearing capacity values calculated based on geological data and different codes are also listed in Table 10-7.

Bearing capacity of testing piles   Table 10-7

Pile number	Static load test		Calculated by geological data	Calculated by undrain shear strength	CPT method		
	Design value	Measured value	Chinese code	US code	Chinese code	US code	European code
SZ1	28000.0	29403.2	28667.0	38685.3	32045.5	37720.1	40232.2
SZ1	13000.0	14824.1	19117.1	25376.7	21401.7	25346.6	28416.3

As Table 10-7 shown, the bearing capacity of the two test piles measured by the static load test meets the design requirements and are closest to the calculation results by geological data. Meanwhile, it can be seen that there are significant differences in the ultimate bearing capacity calculated based on different codes and methods. As far as the CPT method is considered, the bearing capacity calculated using European code is the maximum, and that using Chinese code is the minimum. And the bearing capacity calculated by CPT is greater than that calculated by geological data in Chinese code. The relatively similar bearing capacity are obtained by using the undrained strength method and CPT method according to US code. If we use the bearing capacity calculated by geological data in Chinese code as a benchmark, Table 10-8 presents the bearing capacity comparison calculated by various methods.

Bearing capacity comparison calculated by various methods   Table 10-8

Pile number	Calculated by geological data	Calculated by undrain shear strength	CPT method		
	Chinese code	US code	Chinese code	US code	European code
SZ1	1	1.35	1.12	1.32	1.40
SZ1	1	1.33	1.12	1.33	1.49

As Table 10-8 shown, the bearing capacity of piles calculated by geological data in Chinese code is the most conservative. The bearing capacities calculated by the undrained strength method and CPT method in US code are 30% and 20% larger than that calculated by geological data and CPT method in Chinese code, respectively. The bearing capacities calculated by the undrained strength method and CPT method in European code are 40% and 30% larger than that calculated by geological data and CPT method in Chinese code, respectively.

Questions

10-1   What are the main differences among the Chinese codes, US codes and European codes for the design of foundation engineering?

References

[1] Chen Zhongyi, Ye Shulin. Foundation Engineering [M]. Beijing: China Architecture & Building Press, 1990.

[2] Gao Dazhao. Soil Mechanics and Foundation Engineering [M]. Beijing: China Architecture & Building Press, 1998.

[3] Zhou Jingxing, Li Guangxin, Zhang Jianhong, and etc. Foundation Engineering (Third Edition) [M]. Beijing: Tsinghua University Press, 2000.

[4] Li Guangxin, Zhang Bingyin, Yu Yuzhen. Soil mechanics (Second Edition) [M]. Beijing: Tsinghua University Press, 2013.

[5] Wang Xiequn, Zhang Baohua. Foundation Engineering [M]. Beijing: Peking University Press, 2006.

[6] Yuan Juyun, Lou Xiaoming, Yao Xiaoqing, and etc. Design Principle of Foundation Engineering [M]. Beijing: China Communications Press, 2011.

[7] Zhang Yan, Liu Jinbo. Handbook of Pile Foundation Engineering [M]. Beijing: China Architecture & Building Press, 2009.

[8] Donald P. Coduto. Foundation Design: Principles and Practices [M]. Beijing: China Machine Press, 2004.

[9] Wang Yuqing, Xin Hongbo, Gao Yanping. Earthquake resistance of geotechnical engineering [M]. Beijing: China Water Resources and Hydropower Press, 2013.

[10] Wang Chenghua. Foundation Engineering [M]. Tianjin: Tianjin University Press, 2002.

[11] Miu Lincang, Zhou Yixin, Li Zhihuai, and etc. Design comparison of pile bearing capacity for Chinese and US codes [J]. Journal of China and Foreign Highway, 2016, 36(1):77-80.

[12] National Standard of the People's Republic of China, Code for design of building foundation (GB 50007—2011) [S]. Beijing: China Architecture & Building Press, 2012.

[13] National Standard of the People's Republic of China, Technical code for building pile foundations (JGJ 94—2008) [S]. Beijing: China Architecture & Building Press, 2009.

[14] National Standard of the People's Republic of China, Technical code for ground treatment of buildings (JGJ 79—2012) [S]. Beijing: China Architecture & Building Press, 2013.

[15] National Standard of the People's Republic of China, Code for investigation of geotechnical engineering (GB 50021—2001) [S]. Beijing: China Architecture & Building Press, 2002.

[16] National Standard of the People's Republic of China, Technical specification for retaining and protection of building foundation excavations (JGJ 120—2012) [S]. Beijing: China Architecture & Building Press, 2013.

[17] National Standard of the People's Republic of China, Code for building construction in collapsible loess regions (GB 50025—2004) [S]. Beijing: China Architecture & Building Press, 2005.

[18] National Standard of the People's Republic of China, Technical code for buildings in expansive soil regions (GB 50112—2013) [S]. Beijing: China Architecture & Building Press, 2014.

References

[19] National Standard of the People's Republic of China, Code for seismic design of buildings (GB 50011—2010) [S]. Beijing: China Architecture & Building Press, 2011.

[20] European Committee for Standardization. BS EN 1997-1:2004. Eurocode 7: Geoteehnical Design-Paal: General rules[S]. London: BSI, 2004.

[21] The AASHTO LRFD Bridge Design Specifications, Sixth Edition[S]. American Association of State Highway and Transportation Officials, 2012.